Praise for
The Immortal Mind

"A must-read book not just for anyone grappling with the mind-body problem but also for anyone trying to understand what it truly means to be human." —William A. Dembski, author of *The Design Inference*

"In engaging and easy-to-read prose, Michael Egnor brings his many years of experience with actual patients to back up claims that the human mind is not just an aspect of the brain but a spiritual reality. The reader will come away with faith in his common sense restored by looking at the evidence from science that the authors present. Science, rather than materialistic assumptions current in academia, points to our being persons capable of knowing and choosing freely, which is what the Judeo-Christian tradition maintained all along. An important and timely contribution."
—Fr. Martin Hilbert, The Oratory/Holy Family Church

"*The Immortal Mind* is an important new book advancing thought about that old question: Is there a difference between the mind and the brain? While the topics of the mind and the brain can be complicated, this book is written so well that it is never ponderous or difficult. Stories, including sometimes moving stories of Dr. Egnor's experiences as a neurosurgeon and of his life, as well as clear explanations in addition to the narrative trail, help to make the book readable and compelling. Not easily forgotten, *The Immortal Mind* is a book that continues to provoke contemplation after the last page is read."
—Stephanie West Allen, JD, teacher, writer, conflict-resolution professional, former lawyer and professor

"Based on compelling scientific evidence as well as an unstoppable philosophical argument, *The Immortal Mind* constitutes a masterful demonstration of why the materialist worldview is wrong. In addition, this provocative and inspiring book convincingly explains why we humans are immortal spiritual beings temporarily inhabiting physical bodies during our short time on Earth."

—Mario Beauregard, PhD, neuroscientist, coauthor of *The Spiritual Brain* (with Denyse O'Leary) and *Manifesto for a Post-Materialist Science*, author of *Brain Wars* and *Expanding Reality*.

"An informative and rich exploration of the reality of the human soul's immortality. Neurosurgeon Michael Egnor and journalist Denyse O'Leary present a compelling synthesis of neuroscience and metaphysical inquiry, challenging the dogmas of scientific materialism. This work is not only timely but also stands as a testament to the enduring quest for truth in both science and faith. I highly recommend this book to intellectuals and laypeople interested in the realities about God, objective morality, and the soul's eternal destination."

—Dr. Scott D. G. Ventureyra, author of *On the Origin of Consciousness* and *Making Sense of Nonsense*

"Who better to tell us about the mind than a highly experienced brain surgeon who has performed surgery on literally thousands of brains, probed deep into the brain with electrodes, and treated people with brain injuries? The Immortal Mind makes a compelling scientific case against a purely materialistic view of the human person."

—Nancy Pearcey, professor and scholar in residence at Houston Christian University, author of bestselling books *Total Truth*, *Love Thy Body*, and *The Toxic War on Masculinity*

The Immortal Mind

The Immortal Mind

A Neurosurgeon's Case for the Existence of the Soul

Michael Egnor
and Denyse O'Leary

NEW YORK NASHVILLE

Copyright © 2025 by Michael Egnor and Denyse O'Leary

Cover design by Whitney J. Hicks. Cover art by Shutterstock.
Cover copyright © 2025 by Hachette Book Group, Inc.

Hachette Book Group supports the right to free expression and the value of copyright. The purpose of copyright is to encourage writers and artists to produce the creative works that enrich our culture.

The scanning, uploading, and distribution of this book without permission is a theft of the author's intellectual property. If you would like permission to use material from the book (other than for review purposes), please contact permissions@hbgusa.com. Thank you for your support of the author's rights.

Worthy
Hachette Book Group
1290 Avenue of the Americas, New York, NY 10104
worthypublishing.com
@WorthyPub

First Edition: June 2025

Worthy is a division of Hachette Book Group, Inc. The Worthy name and logo are registered trademarks of Hachette Book Group, Inc.

The publisher is not responsible for websites (or their content) that are not owned by the publisher.

The Hachette Speakers Bureau provides a wide range of authors for speaking events. To find out more, go to hachettespeakersbureau.com or email HachetteSpeakers@hbgusa.com.

Worthy Books may be purchased in bulk for business, educational, or promotional use. For information, please contact your local bookseller or the Hachette Book Group Special Markets Department at special.markets@hbgusa.com.

Print book interior design by Marie Mundaca

All scripture quotations are taken from the Holy Bible, New International Version®. Copyright © 1973, 1978, 1984, 2011 by Biblica, Inc.™ Used by permission of Zondervan. All rights reserved worldwide. www.zondervan.com. The "NIV" and "New International Version" are trademarks registered in the United States Patent and Trademark Office by Biblica, Inc.™

Library of Congress Cataloging-in-Publication Data
Names: Egnor, Michael, author. | O'Leary, Denyse, author.
Title: The immortal mind : a neurosurgeon's case for the existence of the soul / Michael Egnor and Denyse O'Leary.
Description: First edition. | New York : Worthy, 2025. | Includes bibliographical references.
Identifiers: LCCN 2024060389 | ISBN 9781546006350 (hardcover) | ISBN 9781546006374 (ebook)
Subjects: LCSH: Soul. | Neurosciences. | Science and religion.
Classification: LCC BD421 .E38 2025 | DDC 128/.1—dc23/eng/20250307
LC record available at https://lccn.loc.gov/2024060389

ISBNs: 978-1-5460-0635-0 (hardcover), 978-1-5460-0637-4 (ebook)

Printed in the United States of America

LSC-C

Printing 2, 2025

From Michael Egnor:
To my wife, Shelley, the love of my life.
To my children, Daniel, Julie, Emily, and Scott, who make me so proud.
To our Lord and Savior Jesus Christ, Who died and rose so that we might live.

From Denyse O'Leary:
For Elizabeth Dunning, Keith Cassidy, and
Elizabeth Ring-Cassidy, teachers.

Contents

Introduction: Neuroscience from the Chapel Floor 1

CHAPTER 1
The Brain Can Be Split but Not the Mind 13

CHAPTER 2
How Much Brain Does the Mind Need? 36

CHAPTER 3
The Mind Is Hard to Just Put Out 57

CHAPTER 4
When Two Minds Must Share Body Parts 75

CHAPTER 5
The Human Mind Beyond Death 84

CHAPTER 6
The Skeptics' Turn at the Mic 102

Contents

CHAPTER 7
Immortality of the Soul Is a Reasonable Belief 120

CHAPTER 8
Free Will Is a Real and Intrinsic Part of the Soul 135

CHAPTER 9
Models of the Mind—Which One Fits Best? 151

CHAPTER 10
The Human Mind Has No History 168

CHAPTER 11
What Does It All Mean? Neuroscience Meets Philosophy 181

CHAPTER 12
And This All Men Call God 191

CHAPTER 13
Does AI Really Change Everything? Anything? 206

Conclusion: The Truths That Matter Most 219

Acknowledgments 227
Notes 229
About the Authors 259

INTRODUCTION

Neuroscience from the Chapel Floor

WHEN I WAS A medical student, I believed that science could explain everything. I was sure that answers to the big questions—How does consciousness emerge from the brain? Do we have a soul? Do we continue to exist after death?—would be found in the laboratory and the operating room. And so I was surprised when, late one night, I found the answers on the floor of a hospital chapel.

This book is the story of what I learned after I got up off that chapel floor. It is about my search for the evidence that the mind is immortal and that, by extension, the human soul exists.

First, some background. My family was not religious. We believed in God, in a sense. My mother taught me the Lord's Prayer, for example, but we almost never went to church. I was not baptized, probably because my parents never thought of it. We lived in a poor rural area of upstate New York and we were just scraping by. Putting food on the table and clothes on our backs was work enough.

Introduction

My escape from a hardscrabble childhood was science. Science was where facts and truth were found. I was a good student, mostly because I loved science. I was fascinated with the structure of the brain and the motion of the planets and the workings of atoms. In elementary school I memorized the names of the planets and even sent a letter to a publisher when a children's book about astronomy got the number of Jupiter's moons wrong! My cousin was a nurse and she had assisted in brain surgery, so I spent hours asking her about how the brain worked.

What I loved most about science was that it provided objective answers about the world, not mere opinions of the sort that were offered by other ways of investigating things. I dreamed of exploring the planets, discovering laws of nature, and figuring out how the brain worked. In science, all the mysteries of the world—how the universe began, how the brain works, how atoms link together to make molecules—are laid open to explore. I realized that the scientific method—careful, dispassionate collection of data and thoughtful contemplation of what it means—would be my ticket to understanding the grandest mystery of all: how the world works and why we are here to ask the question.

When I was twelve years old, I sat transfixed in my living room and watched the breaking news of the first human heart transplant in Cape Town, South Africa. I thought of it as a miracle—that a dead person's heart could be removed and placed in the chest of a person who was dying of heart disease. My mother's life had been saved by a surgeon—she had a ruptured brain aneurysm when I was a toddler and had lifesaving brain surgery at Columbia-Presbyterian hospital in New York City. In our house, surgeons were heroes who performed delicate, dangerous operations with great skill in order to save lives, including my mother's. To be a heart surgeon or a brain surgeon was the pinnacle of achievement—to be a surgeon was to play a central part in life-and-death drama. That was when I decided I would be a surgeon. It seemed like the most intimate way to get to know what it is to be human and

Introduction

to unravel the mystery of life. I would not merely read books about the heart or the brain; I would see them for myself, and I would learn how to heal people in a hands-on way. Surgery would be a front-row seat to the drama of life!

In 1980, I entered Columbia College of Physicians and Surgeons in New York—the same medical center where my mother's life had been saved by neurosurgeons. Medical school was heady stuff. Like all medical students, I was exhilarated, fascinated, terrified. I fell in love with the anatomy and physiology of the brain. I couldn't put the textbook down. I had been torn between becoming a heart surgeon or a brain surgeon. And there was my neuroanatomy textbook, with its beautiful and intricate drawings of pathways in the brain that let us see and hear and think and feel. That neuroanatomy textbook was the answer to my dilemma. I was going to be a neurosurgeon!

My most important day in medical school came at last. That was the day I first scrubbed for a brain operation—in the same operating suite in which, years before, my mother's surgery was performed!

I was transfixed. The surface of the brain was beautiful—cream-colored swirls of gyri (ridges) and sulci (valleys) with pulsating arteries and veins coursing through the ridges and valleys. This was, I was sure, the seat of the soul (if we had souls!). I wanted to solve the mysteries of the brain and the mind—where our perceptions and thoughts come from, how consciousness comes from this landscape of gyri and sulci, how the mind emerges from this elegant three-pound organ.

At a nursing school dance, I met the love of my life. After we married, we moved to Miami, where I began my training in neurosurgery. It was a grueling but instructive six years. Miami was a huge, violent city, which meant that there was a lot of neurosurgery to be done—brain tumors, aneurysms, gunshot wounds, brain hemorrhages—day and night. It was the hardest work of my life, but I could work in a remarkably intimate way with people with injured and ill brains.

Introduction

I saw what happened when specific parts of the brain are injured, examining and talking with the injured patients during their recovery. I had the privilege of learning firsthand how the mind (I will explain more about this later, but by this I mean the personality, the core of who a person is) changes when the brain is injured. It was a close-up look at the great drama, the mystery of life and of the difference between the mind and the brain.

I was still an atheist at this point. God meant nothing to me. I wasn't an angry atheist—I liked Christians and thought that Christianity was a lovely story—but I also thought it was a fairy tale. Science was my road map to reality—prayer and the Bible seemed to me to be so... *unscientific*. Faith felt like a fairy tale whose origins existed in the brains I was operating on each day. The truth, I thought, was in the operating room and the laboratory in which I spent my days and nights. I was sure that the brain and the body were all there is. I never entered a church, because I feared I would have to leave my brain at the door. Christianity seemed to me to be pure emotion, without contact with scientific evidence or tangible reality. What did a preacher have to teach *me* that could compare to my neuroscience textbooks and my professors in the operating room?

But I Was, Nonetheless, Haunted

I got eerie feelings occasionally—I called them hauntings—that took my breath away. They were disturbing and enticing at the same time. They came mostly when I was alone, with time to let my mind wander, when I was waiting in line or when I awoke during the night and had trouble getting back to sleep. They were the sense that there was a profound truth pressing in on me that I was ignoring, because I was so caught up in the ordinary affairs of everyday life—work and family and leisure—that I was ignoring this elephant in the room. Why was I here? Why does anything exist? What is life all about?

Introduction

The hauntings came unexpectedly when I was alone, or glimpsed a sunset, or woke up at night. In retrospect, I think they came from God, although at the time I believed that they were just momentary "breakthroughs" of the deepest reality into my ordinary life.

Sometimes the hauntings took the form of a recurring daydream—I imagined waking up one morning in a huge mansion, ornate, rich, and beautiful. But I had no memory of how I got there or where I came from, and no knowledge of where I was going. Of course, an urgent phone call or an emergency page on my beeper or a word from my family would quickly interrupt these dreams. As an atheist, I found it easier not to ask myself questions about this mansion. I thought questions could only be explored through science. Medical science was my day job and I believed it was the only reliable way of understanding the world and what happened in it.

So What Changed?

This sense of being haunted got worse when my kids started to arrive. When my first son was born, I looked at his perfect little head and beautiful hair, and thought, "He came from somewhere else, from Someone. He is a gift." I was not only living in a mansion but I was also getting beautiful gifts! Family and friends would say, "He is such a beautiful baby—and he came from you and your wife!" I would always reply, "He came *through* us, not *from* us." I still didn't admit even to myself that I believed in God—but my hauntings were getting more intense. Sometimes I had trouble *not* thinking about them. Why was I alive? Where did I come from? Where am I going? Why had I been given so many wonderful things in life, and why was I given such difficult challenges as well?

But something else was happening, too, and it's something I want to talk about in this book. As I learned more about the brain, I saw

Introduction

many patients with brain damage, whether from birth, bullets, blood clots, or tumors. And yet their minds—their ability to think and reason and believe and desire—didn't seem as damaged as their brains. I had learned in my training that certain parts of the brain control certain abilities—movement, perception, emotion, and memory, for example. But I also saw that other abilities didn't seem to come from the brain at all, or at least not in the same way.

For example, I operated on one woman I'll call Sarah who had a brain tumor in her left frontal lobe. The tumor was invading the regions of her brain that controlled her ability to speak. So I had to remove both the tumor and the brain tissue it had invaded. But I also had to protect the "speech area" in the cortex of her brain. That meant she would need to be awake during the surgery, so I could map the surface of her brain to ensure that her speech area was not being damaged. As frightening as it sounds, this *awake* brain surgery is not painful because the brain itself feels no pain and local injections of anesthetic are sufficient to numb the scalp.

During the surgery, I spoke with her continuously as I was quietly removing most of her left frontal lobe. She maintained a perfectly normal conversation throughout the surgery. That left me wondering: How *does* the brain relate to the mind? How can I remove such large parts of her brain without any effect on her mind—her thinking and reasoning and believing and desiring—at all? This question is a central focus of our book, and we will explore the answers in the chapters that follow.

Day in and day out I lived in my familiar and busy world of work, family, and ordinary things, but my hauntings continued. There were moments when I felt the chill, when I wondered about the questions that followed me for years. Why was I here? Where was I going? Who made this world?

These questions came to a crisis—the turning point in my life—when my younger son was born.

Introduction

The Serious Questions Were Catching Up with Me

When my son was a few months old, my wife and I noticed that he wasn't smiling or making eye contact with us. He would look at objects with interest, but not at people. We started to face the possibility that he might be autistic. This terrified me—I had always dreaded autism. I knew it would be the worst agony to have a child you love who doesn't know you or love you back. I had nightmares about my son as an older child sitting in a room, alone and rocking back and forth, while his schoolmates played baseball and enjoyed normal childhoods.

We took the child to an autism specialist, but he said it was too soon to be sure. We would have to wait before we could know more. But at nearly six months of age, he was still not responding to us. I found it harder each day to go about my daily tasks because I thought about him all the time.

One night, it all came to a head.

I was called to see a patient at a Catholic hospital in another town. As I was leaving the hospital, I passed the chapel. I thought, "I don't believe in God, but I'll do anything now. I just want my son to know me."

I went into the chapel and knelt before the altar. "God," I said, "I don't know if you exist, but I need help. I am terrified that my son is autistic. It's agony to have a child who will never know or love me."

Then I heard a voice—it was the only time in my life I'd ever heard a voice in my head that was not mine—and the voice said, *But that's what you're doing to Me.*

I collapsed in front of the altar. The voice I heard had only spoken seven words, but I felt like He knew me intimately and had been watching me with love and wisdom all my life and that He knew me better than I knew myself. It was like a curtain was lifted, and the Source of my life was speaking to me directly. My heart burned in me.

Introduction

When I recovered, I prayed, "Lord, I will stop doing it to You. I'm sorry. I won't be autistic to You any longer. Please heal my son, and please heal me." I walked out of the chapel a shaken man, and a different man.

The next morning, I called my local Catholic church (my epiphany had happened in a Catholic chapel, and I sensed that this is where He wanted to meet me) and asked to be baptized.

A few days after my prayer in the chapel, I went home in the evening to my son's six-month birthday celebration. That night, he was behaving like a completely normal child, looking us in the eye, smiling and laughing. I knew I had experienced a miracle. It seemed as if, just as I had stopped ignoring the Lord, He had also allowed my son to see me.

The next Easter, I was baptized, along with my son and other members of my family.

What Is Science?

Since I now believed God is real (we had, after all, had a brief conversation!), I rethought my slavish devotion to using only conventional materialist science to understand the world. What had been haunting me all my life was *the quest for truth*. And when I found that Truth, it changed me. I began to see science in an entirely different light.

Science is the organized study of nature according to the causes of things. This means that the causes in nature—the Big Bang, the laws of physics, the astonishing complexity and purposefulness built into living things that could not have arisen by mere evolutionary mechanisms, the marvelous human ability to reason and dream and desire goodness—need not in themselves be natural things. It's entirely possible that ordinary scientific causes in nature come from *outside* nature. Nature is not a closed system. There is room for God—a *need* for God—in science.

Introduction

The purpose of good science is to follow the evidence where it leads and to pursue the truth about the world without ideological blinders. I had followed the evidence, and it had shown me that the supernatural is real. The inference to God's existence and to creation, design in biology, and the reality of the human soul is compatible with science, *because it's true.*

So I went into the library and the operating room with a new resolve to do neuroscience and neurosurgery without blinders. I reread classic research papers on the relationship between the mind (our perceptions, emotions, memories, thoughts, capacity for reason, our free will, and so on) and the brain (the three-pound organ in our skull that generates electrical signals and neurochemicals). I read up on well-respected theories about how consciousness works, and we will discuss them in this book. But now I read neuroscience without materialist blinders—that is, without the presumption that we are simply flesh-and-blood creatures without spirits or souls.

I also read the work of many of the greatest philosophers—Plato, Aristotle, Thomas Aquinas, Ludwig Wittgenstein—to explore a logical framework for understanding these profound questions about our minds and bodies. With this new insight into the human mind and brain, I began, for the first time really, to understand the truth about the human soul.

My Journey and Our Journey

My own Christian journey began with me facedown on the floor of a chapel. And in many important ways, my scientific journey began there as well. It was only after that moment in the chapel that I began to ask the deeper, important questions, the haunting questions—who are we, where do we come from, and where are we going after we die? And I no longer settled for the stock answers that many scientists (like me, up to that point) had naively accepted for so long. My experience in

Introduction

the chapel that night turned out to be only one instance of my encounter with "thin places," as some theologians call them, where the wall between God and man is so thin that we can see through it, if we know how to look. I have since found many thin places—at Mass, during quiet prayer, on long walks talking with the Lord, and during visits to Rome in St. Peter's Basilica and several other beautiful churches in the Eternal City.

In my journey, I discovered that much of what I thought I knew was wrong. Like many scientists, I had been trained to believe that the soul is a myth and the mind is nothing more than the brain—that is, a physical machine. As we shall see, the general intellectual atmosphere in which neuroscience is done today *entails* the denial of the soul. Like a machine, the brain is supposed to be operated by physical forces beyond its own control. When the brain is damaged, our mind is damaged, we learned, sometimes irretrievably. When the brain breaks down and cannot be fixed, we cease to exist.

Yet the more I investigated, the more I discovered that these supposed findings of science were myths. The real findings point in another direction. As a busy neurosurgeon who has performed over seven thousand brain operations, a medical school professor who has taught young medical students and aspiring neurosurgeons for forty years, and as a researcher who has applied the rigorous methods of science to discoveries about how the brain works, I have come to see that the human soul is real and that human beings are not mere machines made of meat. I have come to see, through practical everyday experience with neurosurgery patients; through working with neuroscientists, doctors, and students; and through devoted study of two thousand years of philosophical discoveries about the human mind and soul, that we human beings are *spiritual*, and not merely physical, creatures, created by God and destined for eternal life.

In the pages that follow, we will explore together some of the

Introduction

most amazing discoveries of modern neuroscience and medicine—discoveries you may never have heard of—and we'll explore how they show that there is overwhelming scientific evidence that human souls exist, that some abilities of the mind are actually something separate from the brain, and that our souls do not cease to live when our bodies die.

You will learn about compelling scientific evidence that our minds continue to exist and function even when our brains are severely damaged. You will learn how the personal identities of loved ones continue to exist even if they are in a coma or have dementia. You will learn how we have free will and can exert control over our physical brains. You will learn how science has uncovered powerful evidence that our minds survive death. Finally, you will learn how science now points to the existence of a cosmic Mind behind the universe.

Join me on a journey that started many years ago that night on the chapel floor.

CHAPTER 1

The Brain Can Be Split but Not the Mind

WHAT ACTUALLY HAPPENS WHEN the brain of a living human being is split in half? It's not a theoretical question. Neurosurgeons have done it many thousands of times. As a neurosurgical resident in the late 1980s, I had to do it on a patient we'll call Sam. Epileptic seizures had plagued him daily, essentially destroying his life.

When I started out in medical school, I was amazed that this split-brain procedure—technically called *corpus callosotomy*—could be done at all. It's radical—we really do split the corpus callosum, the massive bundle of millions of nerve fibers that connects the two hemispheres of the brain, in half.

We only do it when there is no useful alternative. Epileptic seizures, which are random uncontrolled electrical discharges in the brain, can jump from one side of the brain to the other through the fiber bundle, causing many catastrophic convulsions each day. Seizures can be diagnosed by their effect on the body, which includes symptoms like uncontrolled shaking of the muscles, loss of consciousness, abnormal perceptions such as flashes of light or tingling on the skin, strong emotional states, or sudden, even random, memories. They can also be diagnosed by medical technology such as electroencephalography (EEG), which records brain electrical impulses via electrodes on the scalp. Seizures can often be treated successfully with medications, but sometimes that's not enough. Radical surgery is a last resort.

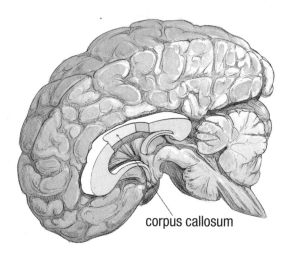

corpus callosum

But surely, I first thought, the trauma involved in cutting the brain in half would be too massive. The effect on the mind would be dramatic and disabling. The corpus callosum's two hundred million axons connect most regions of the brain like a massive telecommunications relay station in the middle of a great city. Cutting through the middle would sever these critical connections and, I supposed, might radically damage the mind—that is, impair the patient's ability to think, to reason, to choose, and even the patient's sense of himself as a unified person. It might even result in two centers of consciousness—the surgery would make one person with one mind into two persons with two minds!

But unexpectedly, the callosum differs from a relay station in one critical way. Despite its size, its critical location, and its myriad connections to nearly all regions of the cerebral cortex, the corpus callosum doesn't seem to have an *irreplaceable* neurological function in the brain. People born without a corpus callosum have lived apparently normal lives—with disconnected hemispheres of their brains![1]

Still, even after I learned that such a surgery could be successful in treating the physical brain, I wondered about what everyone imagines with this kind of surgery—what would cutting the brain in half, slicing right through its usual connections, do to the patient *as a person*? That question goes to the heart of what we are as human beings.

For centuries, we have believed that the brain is the organ of the mind and that consciousness arises wholly from the brain. With the brain cut in half, how would one hemisphere of the brain know what was going on in the other hemisphere? How *could* a person really act as a unified individual with the two halves of the brain disconnected? Well, I would soon get a chance to find out.

I was quite apprehensive, but the six-hour operation went smoothly. We opened a window of bone in Sam's skull, sliced through the membrane that covers the brain, and gently pulled the hemispheres apart, taking great care not to injure the numerous delicate arteries and veins coursing between them or the critical areas on the surface of the brain itself.

Using an operating microscope to get a brilliantly illuminated and magnified view of the deepest recesses of Sam's brain, we gently and methodically cut all of the connections of his corpus callosum from the front to the back, confirming with the microscope that his brain hemispheres had been fully disconnected. Six hours later, we replaced the bone on his skull, sutured up his scalp, and brought him to the recovery room. His vital signs were good.

But was it still Sam?

When I went to see Sam in the intensive care unit the next morning, two small drains channeled bits of bloody fluid from under his scalp, which was bandaged in white gauze. He was drowsy, both from the surgery and from his pain medications.

"How are you feeling?" I asked cautiously.

"Kinda sore."

I told him the good news. We had succeeded in cutting the bundle in half.

As we spoke, I glanced at the EEG machine. Still better news: Sam's brain waves were showing no sign of the seizures that had plagued him daily for most of his life.

Over the next few days, I spoke with him each morning.

"Do you feel like...yourself?"

He seemed perplexed by my question.

"Yes, I do. I feel fine."

Despite his severed brain, he moved his arms and legs normally, his vision was normal, and so was his speech. In fact, the only thing different about Sam—aside from the bandage swathing his head—was that he was no longer having major seizures. And he certainly seemed like a single, unified person.

Over the next few months, I saw Sam several times in the outpatient clinic. That meant I had more time to talk with him and test him. So far as I was able to test him during an office visit, he had normal use of his arms and legs. His sensations were likewise normal—and his personality appeared unchanged.

But with his brain cut in half, what was going on inside Sam's mind? He told me repeatedly that his inner experiences were just the same as they were before the surgery. He experienced life as one person with a clear, single train of thought. He knew that surgery had permanently split his brain into two separate brains, but the only effect he actually noticed was that his seizures were gone.

I naturally worried about Sam because he was *my* patient. But his outcome wasn't unusual among patients who have undergone this procedure. Patients typically reported feeling well after corpus callosotomy, with fewer and milder seizures. They denied any sense of "split consciousness"—that is, feeling like they were two different people, as some neuroscientists have suggested they might. A few patients did experience unusual conditions such as *alien hand syndrome*, in which one hand (usually the left) occasionally moves seemingly on its own. But these side effects were relatively rare and generally temporary.

Since I operated on Sam, I've cared for a number of patients who had split-brain surgery. Like other neurosurgeons, I found no evidence, either from ordinary clinical examination or events in their lives, that indicates their minds were split, even though their brains were.

Epilepsy as a Doorway to the Human Mind

Epileptic seizures have plagued humans throughout recorded history. In the fourth century BC, Hippocrates—perhaps the first physician in the Western world—described seizures in his writings. He recognized that the brain has an intimate relationship with the mind and that epilepsy arises from the brain when it is not functioning normally.

That insight was remarkable for its day. Among his contemporaries, Hippocrates was almost alone in seeing epilepsy as a brain problem. Other proposed explanations were mostly superstitious and unhelpful, attributing the disorder to demons, gods, or the troubles of the soul.[2] In the early nineteenth century, physicians began to investigate epilepsy specifically in relation to brain function.[3] The first known operation on the brain to treat an abscess in an epileptic patient was performed in 1831 (before anesthesia or antiseptic surgery!).[4] But surgical treatment only became a disciplined science in the mid-twentieth century.

This later research vindicated Hippocrates: Epilepsy is a result of a brain that is not functioning normally. We don't often think of the brain as an electrical organ, but that's what it is. Neurons generate currents that transmit signals down their axons to other neurons via chemical neurotransmitters. Normally, our electrical symphony coordinates beautifully to move an arm, for example, or remember an old friend. But sometimes, due to injury, stroke, a tumor, or for reasons we don't yet understand, the neurons begin to generate massive electrical currents instead—and that is what we call a seizure.

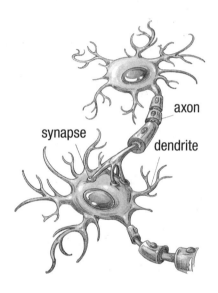

Many seizures begin as tiny aberrant discharges in a few neurons. But then, after a short delay, they spread like a wildfire via the corpus callosum, engulfing the entire brain in *grand mal* seizures. The person loses consciousness, falls to the ground, and convulses in frightening muscle spasms ranging across the body. Although most patients with epilepsy have seizures infrequently, a small number have as many as twenty or thirty seizures a day. Unfortunately, antiseizure medications are ineffective for some patients.

In the early part of the twentieth century, researchers had tested a daring hypothesis on animals such as dogs that had epileptic seizures:

If the problem is coming from the corpus callosum, what if we just *sever* it? Will the animal survive? If so, will seizures diminish? They tried it, and it worked. Post-surgery, the animals generally survived and behaved normally, with less frequent and less severe seizures. By the 1940s, neurosurgeons began performing corpus callosotomy on humans who were suffering severe, intractable seizures, and it helped thousands to live a better life.

What a Split Brain Shows Us About the Mind

I am hardly the only specialist interested in the mental effects of split-brain surgery. This surgery led to truly remarkable neuroscience research in the twentieth century. Here's the most radical thing it tells us: Even when the brain is split in half, many important aspects of the mind remain unified. Thus, *the mind is something that the brain isn't.*

The vast majority of medical and psychology textbooks don't present the mind that way. They treat it as simply what the brain does. In 1994, Nobel laureate Francis Crick, co-discoverer of the structure of DNA and an eminent neuroscientist, summed up the widespread materialist perspective in the neuroscience community:

> The Astonishing Hypothesis is that "You," your joys and your sorrows, your memories and your ambitions, your sense of identity and free will, are in fact no more than the behaviour of a vast assembly of nerve cells and their associated molecules. As Lewis Carroll's Alice might have phrased it: "You're nothing but a pack of neurons."[5]

As a result, neuroscientists tend to ignore the deeper significance of split-brain research. The focus is on minor handicaps that are routinely compensated for, not on the really significant finding. What is most remarkable about the split-brain evidence is the *unity* of the mind despite splitting of the brain. To understand the reality, unity, and independence of the human mind, we need to explore the deeper significance of modern neuroscience findings.

By the mid-twentieth century, study of the losses suffered by people who had brain injuries or strokes showed that some brain functions are *localized*. For example, language, movement, perception, recall, and the experience of emotions are each controlled by specific brain regions. Another key discovery was that the brain's control of the body is *lateralized*. That is, in nearly all right-handed people and most left-handed people, the left hemisphere controls the ability to speak and to understand speech and written language. The right hemisphere is generally nonverbal, and seems to be involved in understanding spatial relationships and other stimuli that don't involve language.

In addition, the two halves of the brain control opposite sides of the body. The left brain hemisphere that "sees" the right visual field also controls the movement of the right half of the body. The right hemisphere that "sees" the left visual field also controls the movement of the left half of the body. But that doesn't mean that perception or movement is split. When the corpus callosum is intact, the hemispheres are so well integrated that they share information through it in milliseconds and work together seamlessly.

Yet, as Sam's story shows, mostly normal function can also continue after the hemispheres have been split. How is this possible, let alone normal? Surely there is some evidence of the split.

A Clever Plan for Detecting Subtle Abnormalities
Some researchers reasoned that subtle losses had probably been overlooked. After all, the people who were relieved that Sam, for example, was no longer suffering life-destructive seizures were not scanning

him carefully for slight disabilities. In the 1960s, California Institute of Technology neurophysiologist Roger Sperry did a famous series of experiments that detected some abnormalities of perception in patients like Sam. They were just the sort of thing that would not be detected in a routine clinical examination.

In his experiments, Sperry asked his split-brain subjects to sit at a table while he presented objects and words in their visual fields, either to the left or to the right. He found that when he showed a picture of an apple to a subject's left visual field, the subject knew what it was. But a subject whose language center was in the left hemisphere would be unable to say the word "apple." That, Sperry concluded, was because vision in the left visual field is perceived via the right hemisphere, which had been disconnected from the left one.

Such disabilities aren't a serious handicap in everyday life because split-brain patients can unconsciously present objects to the appropriate hemisphere simply by moving their eyes. Sperry's experimental setup had explicitly prevented his subjects from using that fix, and that's when their subtle disability became apparent. His research confirmed the theory that, for humans in general, each hemisphere correlates with specific neurological abilities such as language and spatial perception. For that, he was awarded the Nobel Prize in Physiology or Medicine in 1981.

Splitting the Brain Does Not Split Consciousness

But Sperry had also demonstrated something much deeper. He also showed that the mind functions as a unity even when the brain struggles to catch up. After all, apart from his clever experiments, people with split brains had been living normal lives, compensating for their small handicaps in generally unnoticed ways. A person with a split brain was still—in every way that mattered—one person with one mind.

But What If There Are Two Simultaneous Consciousnesses?

Some researchers have responded to this evidence for the unity of mind despite a split brain by suggesting a radical alternative possibility: What if split-brain patients really have two different consciousnesses? These dual consciousnesses might not be distinguishable by ordinary means if they operate simultaneously. The minor perceptual handicaps that Sperry identified could be their signatures.[6] That explanation would seem much more satisfactory to scientists who believe that the mind is simply the activities of the brain.

Proponents of dual consciousness pointed to alien hand syndrome as evidence. When a hand seems to be doing "its own thing," interfering with the movements of the other hand, they suggested, it's likely being guided by another consciousness.

That theory might inspire some great science fiction, but it's not supported by the evidence. Alien hand syndrome is not unique to split-brain patients; it can follow on a variety of neurological conditions where dual consciousness is not remotely suspected.[7] In addition, split-brain patients show divided functions only when researchers prevent them from using common adaptations like moving their heads, as Sperry did. Otherwise they adapt to their situation and perceive things normally. That's hardly much evidence for dual consciousness, especially when nothing that supports it has emerged from other neuroscience sources. So the idea has largely quietly died.

It's not even clear that it makes sense to say that a person has "split consciousness"—after all, if consciousness were genuinely split, there would be two persons and there would be nothing surprising about two persons having two consciousnesses. The very fact that the patient (singular) experiences effects from both the left and the right hemispheres of the brain speaks to the essential unity of human consciousness.

Splitting Perception Without Splitting Consciousness

Two neuroscientists, the Montreal Neurological Institute's Justine Sergent and the University of Amsterdam's Yaïr Pinto, noticed another interesting thing about split-brain patients: The disabilities that Sperry had identified are *only perceptual*. That is, they involve vision but not thought.

In 1983, Sergent reported her research on a split-brain patient who was presented with conflicting messages in each of the two isolated visual fields. The patient was able to respond to the conflicting information with perfect accuracy with either hand, even though each hand was controlled by a cerebral hemisphere that saw only one of the messages. She concluded that the patient could integrate and resolve the conflicting information even though neither separated part of the brain actually saw it all.[8] The patient's *mind* was more than the brain.

In 1986, Sergent reported on a split-brain patient who could integrate visual information, such as whether two lines were aligned or would meet at an angle greater to or less than 90 degrees. The patient showed an accuracy that was well above chance even though no part of the brain "saw" both images.[9]

Again, in 1987, Sergent also reported research on two split-brain patients to whom she presented partial information in each of the split visual fields. Neither field was given enough information to make a final decision. Again the patients made an appropriate decision in most of the trials. She called their unity of consciousness "perceptual disunity and behavioral unity."[10] Thirty years later, in 2017, neuroscientist Yaïr Pinto and his colleagues[11] concluded, after reviewing decades of split-brain research, that the most accurate summary of the research is this: Split-brain patients have split perception but unified consciousness.

In other words, splitting the hemispheres of the brain splits some, but not all, of the mind. It splits what we perceive with our eyes but not what we understand and reason about. This distinction between brain-dependent perception and brain-independent abstract reasoning shows up again and again in modern neuroscience, as we will see.

Deeper: Mapping the Brain of an Awake Patient

Neurosurgeon Wilder Penfield (1891–1976) started out as a philosophical materialist—that is, he believed that the mind is nothing but the brain. Born in Spokane, Washington, he began his medical career by studying under some of the greatest neuroscientists of the twentieth century, including Sir Charles Sherrington (1857–1952) at Oxford. But then he went on to do something that few colleagues had contemplated. He was among the first to operate on the brain while the patient was fully awake.

Why would he want to do this? He was also among the first to face the problem I met with the patient I described earlier in this chapter: When people's lives are being destroyed by epilepsy, surgery to remove specific parts of the brain may be the last hope. But he needed to be sure which parts he could safely remove. And he was practicing long before most equipment for imaging the brain was invented. Thus, much that we take for granted today was still unknown.

Neurosurgeons did know one thing: Regions of the brain differ in importance. The *eloquent* ones manage critical functions like speech, movement, and vision. Other regions can be removed without great loss. By responding to Penfield's questions, a conscious patient could help him determine which was which.

As I mentioned earlier, the procedure is not quite as gruesome as it sounds. The brain has no pain receptors, so the patient needs only local anesthetics for the scalp and the coverings surrounding it. In case you were wondering, during awake brain surgery the patient's head is securely clamped into a device that prevents accidental movement that might cause harm, and patients are given a mild sedative to make the experience less frightening. Thus the surgery can even be done with children who are old enough to cooperate.

Penfield began by perfecting his craft under another pioneer neurosurgeon, Harvey Cushing in Boston, where he mastered the use of local anesthetics. In 1934, with funding from David Rockefeller and other sources,[12] he founded the Montreal Neurological Institute at McGill University and became its first director. Chiseled into the front wall of the "Neuro," as it came to be called, are the words "The problem of neurology is to understand man himself." There, while performing or directing[13] over eleven hundred brain operations on awake patients, he mapped the surface of the cortex and the deeper structures of the brain.

Patiently Exploring the Brain

The main thrust of Penfield's research and work as a neurosurgeon was the surgical treatment of epilepsy. Some of the brain lesions that caused epilepsy—usually areas of scar tissue or congenitally abnormal brain tissue—were large enough that a surgeon could see them.[14] However, Penfield had at least some brain imaging help in the form of *electrocorticography*:[15] electrodes placed directly on the exposed brain that could help detect subtler damage and electrical abnormalities, which indicated that the seizures originated from a specific spot on the brain.[16]

During a typical eight-hour procedure, Penfield would probe the brain to find the region from which seizures arose. When he found it, he would stimulate the surrounding brain tissue to study its functions. That's when he needed patients to be awake, so they could tell him things during the operation. For example, one woman had reported that she could smell burnt toast just before a violent seizure. During the surgery, Penfield asked her to tell him *when* she smelled burnt toast while he was stimulating parts of her brain. That way he could pinpoint and remove the part from which seizures arose,[17] provided that no vital function would be damaged. As he explained, "To be successful, as well as humane, it was essential for the surgeon to explain each step. Indeed, he must take time to talk before and during the operation. He must, in fact, be the patient's trusted friend."[18]

Each operation entailed hundreds of such stimulations and observations, of which Penfield kept meticulous records. By doing this, he became one of the great scientists of neurosurgery. For example, one thing he and colleagues helped produce—for which many students may thank him—is the *homunculus*.[19] That's the technical name for the typical human brain's weird maps of the body. These maps, one for the sensory and one for the motor area of the human cortex, appear highly distorted. They show our bodies as they look to our muscles and skin sensors, not at all as they look to us in a mirror.[20] The distortion reflects the fact that our brains pay far more attention to a sore on the upper lip, say, than to one on the upper back.

Penfield's groundbreaking surgery and research contributed enormously to our understanding of epilepsy. In his research, he identified a type of epilepsy that turns a patient having a seizure into an automaton, wandering about, confused and aimless, even while carrying out relatively complex activities.

One patient, for example, was an accomplished piano player. On some occasions, while he was playing, a seizure would occur, but it wouldn't interrupt his playing—or perhaps it did, but only slightly. He would continue to play the piano with considerable dexterity despite the seizure, even though he was completely unaware of what he was doing. In a sense, his mind didn't know what his body was doing. Another patient would have a seizure while walking home from work, continuing to walk and thread his way through busy streets without

The Brain Can Be Split but Not the Mind

sensory cortex (homunculus)

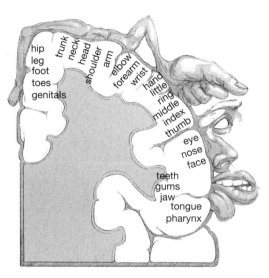

motor cortex (homunculus)

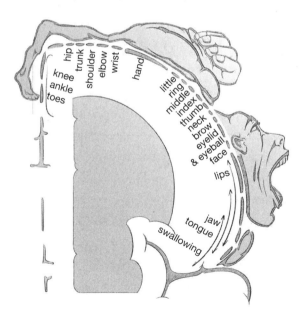

awareness of what he was doing. A third would continue to drive his car during a seizure—and only later realize that he had driven through several red lights!

What Penfield Couldn't Stimulate in the Brain

Penfield was a researcher who asked the deeper questions. For example, how could his patients' brains continue to perform fairly intellectual tasks while they were under the attack of a seizure? And how could they perform these tasks but yet be unaware of them? He reasoned that the mind must still have been engaged, watching and directing, even though the patients' memories weren't recording that.

On later reflection, he likened this "epileptic automation" to computer programming, which was then in its early stages: "The program comes to an electrical computer from without. The same is true of each biological computer. Purpose comes to it from outside its own mechanism."[21] By this, Penfield meant that it seemed to him the brain was merely transmitting the instructions provided by the mind, which was a separate entity from the brain. Reflecting on questions like these, he shed light on the overall mind–brain relationship.[22] That is, Penfield observed that the mind has an existence independent of the brain, and that the mind uses the brain to interact with the world, in a way analogous to the way a computer programmer uses a computer to accomplish tasks.

Slowly, the experiences of epileptic patients began to undermine his earlier, materialist view of the mind. For example, in *Mystery of the Mind* (1975), he recounts, "A young South African patient lying on the operating table exclaimed, when he realized what was happening, that it was astonishing to realize that he was laughing with his cousins on a farm in South Africa, while he was also fully conscious of being in the operating room in Montreal. The mind of the patient was as independent of the reflex action as was the mind of the surgeon who listened and strove to understand."[23]

The Brain Can Be Split but Not the Mind

Penfield had evoked the patient's experience of laughing with his cousins in South Africa by stimulating his brain with an electrode, while at the same time, the patient was fully aware that he was in an operating room. It seemed to Penfield that there were two mental processes at work in the patient simultaneously—the patient's experience of laughter from earlier in his life that was evoked by stimulating his brain with an electrode, and the patient's awareness of his immediate circumstances on the operating table, which was not evoked by Penfield's electrode but seemed to emanate from a source independent of his brain—that is, from the patient's mind. Penfield saw this duality—the dependent "reflex action" of movement or perception or emotions or memories that could be evoked by brain stimulation and the independent mind that could not be evoked by brain stimulation—as a hallmark of consciousness. He saw that there are some mental experiences that do not come from the brain.

Again and again Penfield found a duality in consciousness—some thoughts (such as experiences from childhood) could be evoked by stimulation of the brain itself, while others (awareness of one's current circumstances and the capacity for reflection and reason) could not. For example, in another case, a patient undergoing awake brain surgery suddenly remembered something. Penfield recounted later, "I was more astonished, each time my electrode brought forth such a response. How could it be? This had to do with the mind!"[24] Penfield was amazed that there were some thoughts he could evoke just by stimulating the patient's brain (for instance, a memory), and there were other thoughts that he could not evoke by stimulating the patient's brain (such as the patient's capacity for reason and reflection). The immaterial rational mind that he once thought did not exist—and now realized that he could not evoke—was beginning to claim more and more of his attention as he struggled to understand its relationship to the brain. He admitted in 1975, "The brain has not explained the mind fully."[25]

It may be that a profound personal grief deepened his philosophical reflections. In her forties, Penfield's elder sister Ruth developed a glioma, a tumor of the glial cells that support the structure of the brain. Realizing that her life was in danger, he operated on her himself, removing most of her right frontal lobe during an awake surgery. That was very radical for the early 1930s, but it gave her two more relatively normal years with her husband and six children. When the symptoms returned, however, a second operation by Harvey Cushing proved unsuccessful.

Penfield was deeply affected by Ruth's death, so he wrote about his groundbreaking surgery on her only reluctantly in 1935.[26] Neuroscientist Vaughan Bell notes, "As an academic case study, it is almost unique, as it weaves the medical language of neurology with fragments of memories and heart-felt tributes."[27]

Why Are There No Mind Seizures?

One question in particular came to haunt Penfield: Why are there no *mind* seizures? By "mind," he meant the intellect, the capacity for abstract thought, for reasoning, for introspection. He had observed that epilepsy sufferers could experience seizures while remaining awake. Such seizures involved shaking of the muscles in a particular part of the body, sensations, emotions, and memories. He could produce these same seizures during surgery by stimulating the surface of the brain of an awake patient with an electrical probe. That is, he could cause his patients to move their arms, feel a tingle, see a flash of light, feel an emotion, or have a memory.

For example, after he stimulated a part of the brain during surgery, a mother recalled being in her kitchen listening to her little boy's voice; a man recalled sitting at a baseball game in a small town and watching a child crawl under the fence to join the spectators; another patient recalled being in a concert hall listening to an orchestra. In some cases, the reported memories could be very complex and involve an intricate series of experiences.

But there was a key exception to what Penfield could stimulate in the brain. He was never able to stimulate abstract thought—that is, the sense of self, the capacity to reason, and the exercise of free will.

Natural seizures didn't produce that sort of abstract thought either. Some seizures do involve *forced thinking*. Generally, forced thinking seizures, which are comparatively rare, involve persistent attention to a specific physical object or activity. They have a distinct obsessive-compulsive quality and strong emotional content. At the onset of a seizure, one fifty-year-old woman felt forced to think, "I forgot something that I should do."[28] But what this woman felt was *anxiety*, expressed as the sense of having something to do that she could not remember.[29] She was forced to experience emotions; she was not forced to think about ideas. A 1996 study of three other patients distinguished between these forced thinking seizures and "experiential thoughts," which are reflections on the seizures but not seizures themselves.[30]

Penfield was not saying that seizures involving ideas are uncommon; his point was that they *never* occurred. Neither epilepsy nor neurosurgeons, it seems, could evoke abstract intellectual thought by stimulating the brain. But why might that be?

Abstract Thought Versus Concrete Thought

To understand the relationship between the human mind and the human brain, we need to distinguish between abstract and concrete thought. Abstract thought concerns concepts. A dog owner may think about canine nutrition, which is an abstract concept. Thus the owner reads the ingredients list on a sack of dog food. The dog thinks concretely; he thinks about his next bowl of food.

The Immortal Mind

The distinction between abstract and concrete thought does *not* turn on complexity. An abstract thought may be quite simple, like 2 + 2 = 4. A concrete thought, like navigating a notoriously busy intersection, may be quite complex.

Think about a triangle drawn on a piece of paper. In your imagination, you form a mental picture. It may be any shape of triangle—right-angled, isosceles, or equilateral, or maybe a less common angle. You can imagine its borders as thick or thin, colored or patterned. But it is a particular (concrete) triangle, perhaps like the one you saw in a book yesterday.

Now, think about what a triangle *is*. That is, think abstractly, without actually looking at a triangle or forming a mental image of any specific triangle. Understood abstractly, a triangle is a three-sided closed plane figure with straight sides and with internal angles that sum to 180 degrees. That is a universal definition; it describes all triangles rather than any particular one. It is a remarkable fact that you cannot draw a triangle on paper that is perfect—and yet your abstract understanding of a triangle is always perfect! You understand perfectly well something you have never seen and can never see. That is because your thoughts about the definition of a triangle and about a particular triangle are fundamentally different types of thought.

This classical distinction between abstract and concrete (or particular) thought can also be demonstrated by the *chiliagon*, which is a polygon with a thousand sides. You cannot imagine a particular chiliagon because you can't hold in your imagination any clear picture of a closed figure with that many sides. But you can easily comprehend it abstractly by its definition. Thus, your understanding of a chiliagon is entirely abstract.

It works the same way with language. You can think about the American presidency (abstract) or about John Kennedy (concrete). You can think about slavery (abstract) or about Harriet Tubman (concrete).

Abstract thought in general is *qualitatively* different from concrete (particular) thought. It is a difference in kind, not just in degree.

Abstract Thought and the Difference Between the Mind and the Brain

Penfield was keenly aware of the distinction between abstract and concrete thought. He searched the human brain for the region or location that controls the ability to reason that gives us abstract thought. But during his operations on awake patients, he found that he could stimulate only four capacities—movements, sensations, emotions, and memories. In none of his operations could he stimulate the distinctly human capacities of reasoning and free will. What is more, he found the same thing with natural epileptic seizures—they evoke only movements, sensations, emotions, or memories, but never reasoning or free will. There are no "arithmetic seizures," "logic seizures," or "Shakespeare seizures."

Thus, he concluded that abstract thought is a function of something other than or beyond the physical brain. He came to define the mind as the element in an individual "that feels, perceives, thinks, wills, and especially reasons."[31] His old neuroscience mentor at Oxford, Sir Charles Sherrington, had, as it happens, made a similar journey. He too decided, "Mind, meaning by that thoughts, memory, feelings, reasoning, and so on, is difficult to bring into a class of physical things."[32]

Charles Hendel (1890–1982), a close friend of Penfield's, was a professor of moral philosophy at McGill University. He tried to help him confront the evidence from his neurosurgery practice that so fascinated and troubled him. Hendel wrote to him at one point, "Your autobiographical material *is* powerful, the testimony of your patients is convincing, and your development toward the mystery of the mind is convincing beyond any philosopher's argument. Think it over."[33]

Penfield continued to wrestle with the evidence and described his conclusion in *Mystery of the Mind*, which was published not long before his death:

> Since every man must adopt for himself, without the help of science, his way of life and his personal religion, I have long had my own private beliefs. What a thrill it is, then, to discover that the scientist, too, can legitimately believe in the existence of the spirit![34]

But Is the Human Spirit That Penfield Came to Believe in Immortal?

In his dialogue with his philosopher friend Hendel, Penfield had also cautiously concluded, "The mind must be viewed as a basic element in itself... That is to say, it has a continuing existence."[35] This seems to mean that the mind or human soul does *not* die with the body.

He drew that conclusion from his lifetime of neuroscience research. But we can be confident that the human soul is immortal for several other reasons as well. We'll talk about them in more detail in chapter 7, after we look at the current medical research on near-death experiences. For now, it's worth noting a few of these reasons, to demonstrate that Penfield's widely shared hope is not just wishful thinking.

Contrary to what we sometimes hear, the important question is *not* whether the human soul exists. Of course it exists. The soul of any life-form is its life. In humans, it is the sum of all the activities, conscious and unconscious, that make us living human beings, not dead ones. The human soul animates the human body just as a sparrow's soul animates a sparrow's body. But most humans, including very wise ones, have believed that the human soul survives the death of the body because it has an immortal aspect as well as a mortal one.

The immortal aspect of the human soul—that is, the mind that Penfield observed while he was cutting into the brain—is a unity. It has no parts, so it cannot be split or multiplied. Not even when the brain is split in half. As we will see in the next few chapters, even when the brain is largely absent or shut down, the human person we discern is present in whole, not in part. And even when two souls must share parts of a body, as with conjoined twins, they remain whole, separate, and distinct persons.

Because the soul is not composed of separate parts, it cannot *de*compose, as a dead body does. Like the abstractions it can uniquely comprehend, it is not mortal. The number 127 is not going to die; neither will the mind that comprehends it. So we should not be surprised by the evidence we will encounter from verified near-death experiences.

For now, let's look at what may happen when a whole human mind must work with a deficient brain.

CHAPTER 2

How Much Brain Does the Mind Need?

EARLY IN MY NEUROSURGICAL career, I was called to the emergency room to examine a four-year-old boy named Charles. He had been playing on a sofa and fell on his head, twisting his neck. He got up, told his mom that his head hurt, and then lapsed into a coma.

The CT scan showed that he had torn his vital vertebral artery, which traverses the bones of the neck and provides blood flow to critical parts of the brain. His brain was severely damaged and swelling dangerously. Quite simply, Charles was dying.

We rushed him to the operating room, where I removed the permanently damaged part of his brain—half of his cerebellum—in order to take the pressure off his brain stem and save his life. The cerebellum plays an important role in coordination, so I faced the unfortunate task of telling his family that he might be handicapped, both from the stroke and from the loss of the damaged part of his brain.

Fortunately, Charles surprised me; he survived and recovered quite well. In fact, several weeks later, during a follow-up visit in my office,

How Much Brain Does the Mind Need?

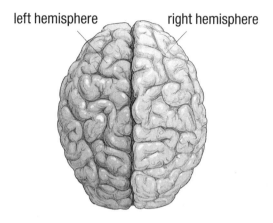

I could find no neurological deficit at all. His cognition, speech, and motor skills were like that of any other four-year-old, even though he was missing half of his cerebellum.

Years later I heard from Charles, after he had grown up and gone on to college. He sent me a photo of himself, dressed in his NCAA college basketball uniform. He was a superb athlete, playing basketball at a high level, and he was even dreaming of a professional career—with almost half of the part of his brain controlling his coordination missing!

Here's the conundrum in a nutshell: Every neuroanatomy and neurophysiology textbook I studied as a medical student described the function of the cerebellum in terms of circuits, neural networks, and computation. And so according to these textbooks, it's highly unlikely—if not impossible—that anyone could sustain even a tiny injury in such a complex "computer" and still retain normal coordination. Let alone the superb coordination that allowed him to play a sport at a high collegiate level. Yet it happens. Neurosurgeons routinely remove large parts of the cerebellum to treat strokes and tumors, and most patients retain normal coordination.

It doesn't happen only with the cerebellum either.

It was only a few years before that, when I was a resident in training, that I was performing brain surgery on Sarah. As I noted in the introduction, she had a tumor near the border of the area that controls speech—Broca's area—and it was infiltrating her brain. That meant removing large areas of her left frontal lobe that were intermingled with the tumor. But I had to know its *exact* location in order to remove as much of the tumor as possible without damaging her Broca's area and leaving her unable to speak.

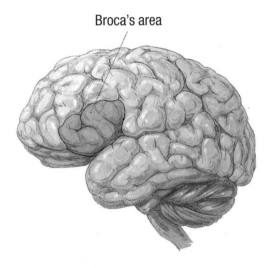

To "map" Sarah's speech area, I needed to keep her talking continuously, while gently and methodically stimulating her cortex with an electrical probe in various locations—much as Wilder Penfield had done over fifty years earlier. As we chatted, the gentle stimulation of the probe would interrupt the electrical signals in her brain for a few moments. That caused speech arrest—a brief pause in her conversation. Those moments when she was briefly unable to speak told me that the brain area I was stimulating was vital for speech and needed protection. Like Penfield, I used only mild local sedation, numbing up her skin and skull, but kept her fully conscious otherwise.

Mostly we talked about the weather, her family, how she was feeling during this stressful time, and—a reliable topic—what she thought of the hospital food. While I was removing the bulk of the tumor, and a large part of the left frontal lobe of her brain along with it, I was startled by her question from under the surgical drapes: "Doctor, what's that sound?"

"Just the sound of the instruments," I replied.

I wasn't being entirely candid. That *sound* was a good portion of her frontal lobe going up my sucker into a canister. It is best not to be too explicit at such a moment.

"It's loud," she commented, half laughing from nervousness and a sedative. "How's the operation going?" She was drowsy, but quite coherent.

I told Sarah that the operation was going well and we continued to chat as I worked. Her brief episodes of speech arrest when I probed her speech area had enabled me to protect it, and she went on to recover nicely. Her tumor was found to be relatively benign—a low-grade astrocytoma—and her prognosis was good.[1]

What Parts of the Brain Are Absolutely Necessary for Normal Function?

As we saw in chapter 1, the brain is lateralized; the two halves do different jobs. Removal of or injury to some areas of the brain can cause profound disabilities. Yet it is also true that taking out key parts—in some cases even up to half—of the brain doesn't necessarily affect the mind. In fact, much of the art and science of neurosurgery is learning to distinguish between vital areas ("eloquent brain") and nonvital areas ("non-eloquent brain").

At the Renaissance School of Medicine in Stony Brook, New York, I am the residency training director for young neurosurgeons. We spend a lot of time discussing which areas can safely be removed if absolutely necessary, and which must be protected at all costs. When I was in residency training myself, one of my professors used to say, "The art of neurosurgery is knowing what you can get away with."

Areas of the brain that can safely be removed (if absolutely necessary) include major areas of the anterior frontal lobes and some areas of the parietal lobes, some areas of the occipital lobes, most of the temporal lobes, and most of the cerebellum. In fact, a common way to gain access to the deep middle parts of the brain is to start by removing a core of brain tissue through the parietal lobe.

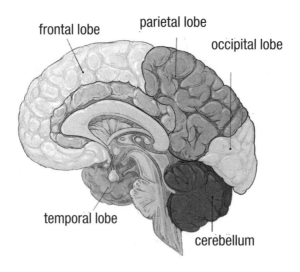

Areas of the brain that can never be safely removed include the left inferior lateral parts of the frontal lobe and inferior anterior parts of the parietal lobes (which are called Broca's area and Wernicke's area, respectively); the posterior part of the frontal lobes (which controls movements of the body); the anterior parts of the parietal lobes (which control sensations from the skin); the medial parts of both of the temporal lobes (the hippocampi, which are essential for memory—one, but not both, can be safely removed); the back parts of the occipital lobes (which control vision); the anterior parts of the cerebellum near the brain stem; all of the deep brain structures such as the basal ganglia (which control movement); the thalamus (which is involved in movement, sensation, and consciousness); the hypothalamus (which controls vital functions such as temperature regulation, hormones, and fluid balance in the body); and all of the brain stem and spinal cord.

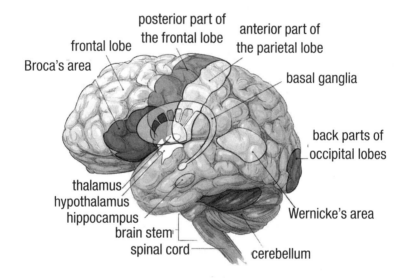

Lastly, some areas of the brain, such as the optic nerves and the stalk of the pituitary gland, are dangerous even to *touch*.

This dichotomy—some very large areas of the brain are quite safe to remove, while some areas must be protected at all costs—was surprising to me in my early years of training. Again, it did not fit with what I had learned in my textbooks. But the distinction between eloquent and non-eloquent parts of the brain mirrors a critical distinction between aspects of the mind. And that's where things get really interesting.

The Immortal Mind

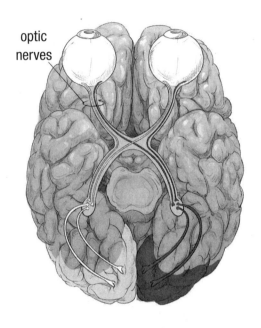

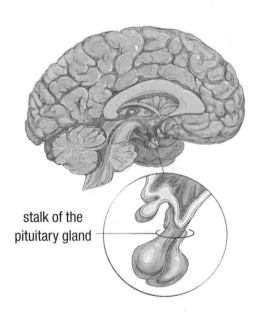

Does the Mind Have a Specific Place in the Brain?

As we can see from the diagrams, four of the very distinct activities of the mind—perception, movement, memories, and emotions—are generated and controlled by specific regions (*loci*) in the brain. Other activities of the mind, such as intellect (the capacity for reason and abstract thought) and free will, don't appear to be generated by such well-localized regions. That is, intellect and will don't *map* to the brain in the way that perception, movement, memory, and emotion do.

For example, when I write a mathematical equation on a piece of paper, the area of my brain that generates the movement of my right hand can be localized to millimeter accuracy on the precentral gyrus of my left frontal lobe. But my *understanding* of the equation cannot be localized at all. Likewise, if you are adding your signature to a letter about an important public issue, I can point to the same tiny area in your brain that guides your hand. But no one can pinpoint where, exactly, you decided to sign the letter of concern.

In short, part of the mind (initiation of movement or the capacity to form new memories, for example) maps onto specific parts of the brain. But part of the mind (*understanding* ideas) doesn't. Of course, if a brain is seriously damaged, intellect and will can be impaired. But

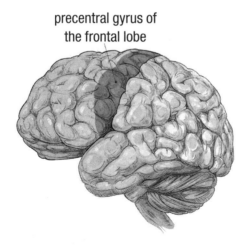

that is because the brain damage impairs memory or perception areas that are needed for the mind to understand what has happened or is happening. In any event, there are many documented cases of people continuing to think normally despite losing or missing large parts of their brains.

Life Without Key Brain Parts

Game of Thrones star Emilia Clarke (playing Daenerys Targaryen) had large parts of her brain removed due to life-threatening aneurysms in 2011 and 2013. She told media in 2022, "The amount of my brain that is no longer usable—it's remarkable that I am able to speak, sometimes articulately, and live my life completely normally with absolutely no repercussions. I am in the really, really, really small minority of people that can survive that."[2]

Understandably, Clarke felt pretty much alone. But her situation is actually not so unique. In 2023, National Public Radio profiled Mora Leeb, whose entire left hemisphere was surgically removed when she was nine months old due to intractable epilepsy (she was having hundreds of seizures per day). But when she was interviewed at fifteen, apart from some limitations in speech and movement, she was a typical teenager. NPR journalist Jon Hamilton recounts, "Mora began by telling me a joke: 'How do you make a hot dog stand?' she asks. 'You take away its chair.'"[3] As Hamilton notes, puns require "a fairly sophisticated understanding of language."

Emilia Clarke and Mora Leeb defy the widely accepted view that human thought is controlled entirely by the physical brain. But they are not alone; medical literature offers a striking number of such cases. For example, a 2019 study of six people who had up to half of their brains removed (hemispherectomy) showed that, despite the loss, each of them adapted well.[4] That is, they live without disabling neurological problems. If you were to meet them (and you never know, you may have met one of them!), you would not notice anything unusual. Similarly, a 2022 study of forty adults who had half the brain removed as children to combat severe epilepsy found that despite that, they performed within 10 percent of other study subjects on face and word recognition.[5]

Medical researchers who study such cases do not offer a simple explanation. In the case of one four-year-old boy, UD (a pseudonym), who had one-sixth of his brain removed but functioned quite well, the researchers admitted, "We do not yet have a definitive account of the exact neurobiological mechanisms that trigger and drive plasticity, nor do we have clear specifics on how the altered architecture is implemented."[6] In short, our current model of the mind–brain relationship doesn't help us understand why he and others have suffered as little damage as they did from as much loss as they endured.

But now what if careless nature happens to leave out large parts of the brain?

What's Missing Is Not the Whole Story of a Person

I had known what to expect when Katie was born, along with a twin sister, but that didn't make it easier when the results from the MRI came in. The neuroradiologists hadn't had a chance to read the results yet, but as I scrolled up and down the images on the computer in my office, the message was obvious. Her brain was, for the most part, *missing*.

There was a sliver of frontal lobes and a plate of deep brain tissue and brain stem. Her parietal, temporal, and occipital lobes were bowed down, as if they were carrying on their backs the spinal fluid (which is essentially water) that mostly filled her skull.

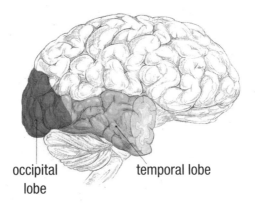

If you were to look at Katie from the outside, she looked like a normal newborn. But inside, her little skull was full of water, except for a rim of brain tissue along its base. This wasn't even entirely hydrocephalus—water on the brain. This was water *in place of* her brain. So while Katie looked like a normal newborn, she seemed to have little chance at a normal life like that of her fraternal twin lying in the incubator next to hers.

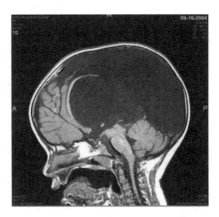

I dreaded talking with her family. True, we had seen the problem on her prenatal ultrasound. I had met with her mom and dad and explained that their daughter would be born with only a small fraction of her brain. They expected her to be disabled in a profound way. But I encouraged a bit of hope. I had been a pediatric neurosurgeon for more than a decade by then, and I knew that the structure of the brain on ultrasound imaging is *not* the whole story of a person. We would have to wait and see. And now, after seeing her MRI, we knew more—but it was not encouraging.

Katie would probably need surgery. Children with large amounts of spinal fluid on the brain usually need a shunt, a small plastic tube inserted into the brain and tunneled under the skin to drain fluid into the abdomen. It would prevent the massive accumulation of water that, if unchecked, would cause her head to expand like a balloon over a few days and eventually kill her.

The shunt would keep her alive, but most of her brain was still missing. I left my office and took the elevator to the newborn intensive care unit. Katie was nestled asleep in her bassinet, which surprised me. Ordinarily a baby with such a profound malformation of the brain would still be in an incubator, as a sort of external womb. But there she was, in a bassinet, wrapped in a blanket, not hooked to machines.

I spoke with her intensive care doctor. He told me that she wasn't sick and didn't need the machines. I measured Katie's head—36 centimeters around, which is normal for a newborn. The water in her skull wasn't causing excessive pressure on her brain. That was reassuring, a small victory. She wouldn't need a shunt on her first day of postnatal life. It turned out, to my surprise, that she never needed surgery.

I checked up on her over the course of the next few days and weeks, and each time I examined her, she displayed normal growth and development.

Katie grew up to be a bright, happy child, and I see her for a routine examination every year. At every stage of her life, she's excelled. She sat and talked and walked earlier than her twin sister. And she made the honor roll in high school.[7]

I've treated and cared for scores of kids who grow up with brains that are deficient and minds that thrive. It is entirely true that some people who are born with large parts of their brains missing are profoundly handicapped. But surely the remarkable thing is that *so many are not.*

Suppose everybody who lacked a certain part of the brain was disabled in a specific way. The brain would be like a machine. Take out a part and it just doesn't work or work right. Then the idea that the mind is merely the physical functions of the brain—the materialist view—would be hard to argue with. But that's not at all what we see.

Brain Imaging Shows That the Brain Is Definitely Not a Machine

Few neuroscientists would have predicted what we discovered when we began imaging brains routinely for medical diagnosis: that some people function normally throughout life despite the fact that there is nothing but water (cerebrospinal fluid) where large parts of the brain should be.

Here are two examples from my own practice: Maria had only two-thirds of a brain. She needed a couple of operations to drain fluid, but she thrives. She finished her master's degree in English literature and is a published musician. Jesse was born with a head shaped like a football and half full of water. Doctors told his mother to let him die at birth. She disobeyed, and Jesse became a normal happy middle schooler who loved sports and wore his hair long.

These cases are unusual, yes, but not unique. According to a recent study at the University of Geneva, one in four thousand people is born without the corpus callosum that joins the two halves of the brain (and must sometimes be surgically split). Twenty-five percent of the people who have nothing but water in that space show no symptoms. About half have some brain-related problems, and 25 percent have serious ones.[8] But, as with the surgical split-brain patients, if the mind is nothing but the activities of the brain, how could any of them show *no* symptoms?

People who are missing some brain parts due to natural causes can even be gifted in the very areas that we would expect to be present and fully functioning. Helen Santoro, who lacks a left temporal lobe, wrote about it for the *New York Times* in 2022:

> They told [my mother] I would never speak and would need to be institutionalized. The neurologist brought her arms up to her chest and contorted her wrists to illustrate the physical disability I would be likely to develop.[9]

Santoro's parents enrolled her in a project that tracked the progress of children with disabilities. However, in addition to meeting conventional development milestones, she showed a passion for the very language-related activities for which a deficiency had been predicted. At fifteen, she was dropped from the disability study as a result. But by then she had already decided to major in neuroscience. One thing she learned was that the brain is not as rigidly compartmentalized as some analysts have thought.[10]

Other adaptations to natural brain absences are surely just as remarkable. In 2014, a twenty-four-year-old Chinese woman in Shandong province complained about dizziness and nausea. A CT scan showed that she did not have a cerebellum, the part of the brain underneath the two hemispheres that coordinates movement and balance.[11] Though as a child she had been a very slow walker and talker, and she has continuing bouts of clumsiness, she had finished school, got married, had a child without complications, and was living a normal life at the time of her scan.[12] The researchers studying her case found nine other living persons identified in the medical literature from 1982 onward as having no cerebellum.[13]

Of these, one was found to be completely normal at fifty-eight years and two others were mentally normal, with some physical issues.

Similarly, one girl, assessed by researchers at fourteen, was born without an entire left hemisphere (hemihydranencephaly). A research team that evaluated her mental abilities from the age of fourteen months through fourteen years reported in 2020 that her "spatial, number, and reasoning skills" were average to above average.[14] Neuroimaging data suggested that her right hemisphere was enabling these skills on its own. While her condition is rare, it is not unique. In 2013, a different research team identified nine such cases recorded since 1971, concluding that "destruction of one hemisphere may be not always associated with severe neurologic impairment and may allow an almost normal life."[15]

Lastly, some people have survived natural brain absences that were even more severe than this. A forty-four-year-old French civil servant who had led a normal life (married with two children) went to a hospital complaining of mild weakness in his left leg. Researchers learned that he had had hydrocephalus as a child and decided to investigate further. He was found, using a CT scan and an MRI, to be apparently missing 50 to 75 percent of his brain: A large chamber filled with fluid left room for only a very compressed thin sheet of brain tissue. The researchers reported in 2007, "On neuropsychological testing, he proved to have an intelligence quotient (IQ) of 75: his verbal IQ was 84, and his performance IQ 70." That's somewhat on the low side but it is not classed as a level of disability. And he was, after all, a working adult with a family.[16]

Is Neuroplasticity a Sufficient Explanation?

Max Muenke, who studies pediatric brain defects at the National Human Genome Research Institute in Maryland, says, "If something happens very slowly over quite some time, maybe over decades, the different parts of the brain take up functions that would normally be done by the part that is pushed to the side."[17] That's neuroplasticity. And, taken by itself, it cannot be a complete answer.

If a child is born without eyesight, good hearing becomes especially important. But the child's ears do not just start seeing. The mind, remarkably, seems to try to work with whatever brain is available, to do things for which that part of the brain was not originally adapted.

Here's the conundrum that these cases pose: On the one hand, a neurosurgeon must be very careful about removing brain tissue that is doing something important. On the other hand, people with very dramatic natural brain losses or absences can have normally functioning minds. If the mind were merely the activities of the brain, that would not be happening. Neuroplasticity is quite real, but it is not magic. A deeper reality underlies it.

Neuroscientist and science writer Tom Stafford cautions that we really don't know as much about the brain as we might imagine:

> We don't often shout about it, but there are large gaps in even our basic understanding of the brain. We can't agree on the function of even some of the most important brain regions, such as the cerebellum... Every so often someone walks into a hospital and their brain scan reveals the startling differences we can have inside our heads. Startling differences which may have only small observable effects on our behaviour.[18]

Despite what we don't yet know, one thing we can say for sure is that the popular image of the brain as a "meat computer" is wrong. It is not like a machine at all.[19] Through all this, the human mind remains not only a unity but also an agency—it can *do* things—even while working around very severe natural brain deficiencies.

But What If Both Hemispheres Are Missing?

Some brain deficiencies are even more extreme than the ones we've described so far. Some children are born without *both* hemispheres, a condition called *hydranencephaly*. At one time, this condition was diagnosed when a doctor shone a flashlight right through the baby's head. With no hemispheres to impede it, the light passed right through the water (cerebrospinal fluid).

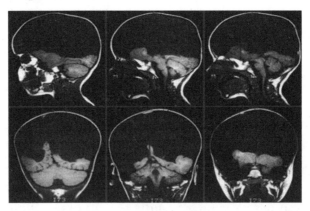

The cause is usually a massive intrauterine stroke that destroys the whole brain above the brain stem, leaving only the brain stem, the basal ganglia, and the meninges (membranes).[20] As a pediatric neurosurgeon, I have cared for some of these children. Although they are quite handicapped, they are certainly conscious and interactive, with a full range of emotions—laughter, crying, glee, fear, and such. In fact, today the condition is often not even diagnosed until several months after birth when the child fails to meet development milestones, thus prompting a neurological examination.[21] Complete destruction of the cerebral hemispheres appears to be fully compatible with basic sentient consciousness.

How Much Brain Does the Mind Need?

Hydranencephaly is rare and not widely studied, so there is not a large amount of clinical research from which to draw conclusions. We do know that such children have a short life expectancy. The medical literature records survival up to nineteen years[22] and twenty-two years.[23] However, one enterprising neuroscientist found a way to have a closer, on-the-ground look at their lives. In 2004, Swedish neuroscientist Björn Merker, preparing a research paper, spent a week at Disney World with five families who each had a child with hydranencephaly. The children, ranging in age from ten months through five years, showed considerable basic awareness: "These children are not only awake and often alert, but show responsiveness to their surroundings in the form of emotional or orienting reactions to environmental events."[24]

As science writer Bruce Bower explained, "They reacted to what happened around them and expressed a palette of emotions. A 3-year-old girl's mouth opened wide and her face glowed with a mix of joy and excitement when her parents placed her baby brother in her arms."[25] Similarly, Merker observed that at home, the children recognized familiar faces and preferred some situations, toys, tunes, or videos over others. Surprisingly, their hearing was generally good despite the lack of an auditory cortex. Merker concluded from his findings that the brain stem supports a basic form of conscious thought.[26]

In 1999, pediatric neurologist D. Alan Shewmon and colleagues published a report on four children between five and seventeen years old who showed "total or near-total absence of cerebral cortex." The subjects nonetheless showed some awareness, "for example, distinguishing familiar from unfamiliar people and environments, social interaction, functional vision, orienting, musical preferences, appropriate affective responses, and associative learning."[27] While no one disputes that these children have serious handicaps, Shewmon's team concluded, among other things, that their development may be hampered by a "self-fulfilling prophecy." We don't try what we are sure will fail—but that's not experience; it's just assumption.

The Dilemma Consciousness Poses

Consciousness in hydranencephalic children raises some interesting questions about human consciousness in general. Even though it is central to our human experience, it's a nebulous concept, devilishly difficult to define. For example, we are conscious in our dreams, yet "unconscious" when we are asleep.

Most of the current theories popular among neuroscientists[28] propose that consciousness is the by-product of the processing of neurons in the cerebral cortex. Yet that's the part of the brain that is completely missing in quite conscious children with hydranencephaly. So here's the dilemma: Although the cerebral cortex is considered by researchers to be the "thinking" part of the brain, some basic thinking occurs *with no cortex at all*.

Pointing to such children as an example, University of Cape Town neuropsychologist Mark Solms has instead made a case for a close relationship between human consciousness and the brain *stem*, the "primitive" part of the brain that we share with vertebrates generally, including goldfish. For instance, he noted on a 2021 podcast with me, if even a small part of the brain stem's *reticular formation*—a network of neurons in the brain stem—is damaged, consciousness disappears. But it's different with kids with no brain hemispheres. "These kids are conscious. There is something that is like to be them, that they have qualitative experience, there's a content to their minds, and yet they have no cortex at all."[29]

When Solms says that there is something *that it is like* to be them, he is applying philosopher Thomas Nagel's famous and widely accepted way of understanding consciousness: If there is something that it *is like*

How Much Brain Does the Mind Need?

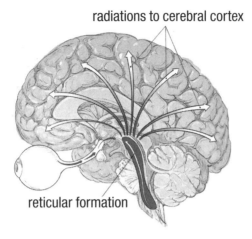

to be you, you are conscious.[30] Using his criterion, consciousness can be usefully distinguished from intellectual ability.

Consciousness can also be distinguished from arousal. Arousal is a state of alertness and heightened interaction with our environment, which arises from the reticular activating system. As we will see in the next chapter, however, people in the deepest level of coma, who show no sign of arousal whatever, may be quite conscious. When brain-scanning equipment is used to communicate, they may be quite capable of understanding others and even answering complex questions. We agree with Solms that arousal is generated by the brain stem; consciousness, on the other hand, is a much more slippery concept. It's not even clear how we *can* define it, let alone explain it. The whole research area has become highly controversial. In 2018, a reporter for the *Chronicle of Higher Education*, attending an academic conference on the topic, found himself asking, "Has Consciousness Lost Its Mind?"[31] In 2023, charges of "pseudoscience" were levied by over a hundred neuroscientists and philosophers—many of them prominent worldwide—against the *leading* theory of consciousness![32] More on that later. At any rate, there is no evidence that consciousness resides in any specific area of the brain.

What the Neuroscience Tells Us So Far

So far, neuroscience has shown us that the human mind is a unity. It can adapt not only to a surgically split brain but also to a variety of brain absences, even radical ones. But now let's range a little further: Can the human mind function when the whole brain is deeply comatose? What about when the brain is afflicted with senile dementia or dying? Let's look at the remarkable and unexpected things neuroscience can tell us about that.

CHAPTER 3

The Mind Is Hard to Just Put Out

WHEN I WALKED INTO Beth's room in my clinic, I was running a bit behind, and I hate to keep patients waiting. So I expected this visit to be brief—just a quick recap of her progress. Beth had lapsed into a coma following her brain hemorrhage. We had succeeded in surgically removing the blood clot, and after a couple of weeks of coma she was awake and had been in rehab for several months. She had been making a very good recovery. Three months out from her surgery, she had a normal neurological exam.

Sometimes the brain tolerates bleeding and surgery quite well, especially if the hemorrhage is in a relatively safe area. Beth told me that she was pleased with her progress. I reminded her that she had been through a lot—a brain hemorrhage, catheters stuck in all sorts of places in her body, a breathing machine for ten days, and major brain surgery. I quipped that, of course, she was unaware of any of this—after all, she was comatose. It was really her family who had the toughest time, watching all of these things happen to her and worrying about her prognosis. *They* had to go through all of this fully awake!

"I was awake too, kind of, at times," she replied. *Really?* I had heard patients say this before. I asked her what she meant.

"I remember hearing things. I couldn't see, but I remember some conversations in the room, about my condition, my medications, my vital signs, things like that."

As I left, I made a mental note to remind my team in the intensive care unit to always be careful what they say in the room of a comatose patient. We often think that because they are in deep coma and hooked up to an imposing array of machines, they don't know what's going on around them.

Nurses, however, and old neurosurgeons like me, know that comatose people sometimes *are* aware of their environment. That is why it is a maxim in intensive care nursing to never say anything in the room of a comatose patient that you wouldn't say to that person wide awake. Patients sometimes even recount what they heard while in coma after they wake up. I have a colleague who was chastised by a formerly comatose patient for telling a joke in her room that was in poor taste!

A more serious problem is that patients' vital signs can change dramatically in response to frightening conversations around their beds. Discussion of a poor prognosis or the need for further surgery can send a comatose patient's blood pressure and heart rate skyrocketing. I've had some patients for whom gentle optimistic comments in the ICU room were clearly therapeutic—their blood pressure would return to normal and the swelling in their brains would subside more quickly. That's why I always tell families of comatose patients to assume that their loved ones can hear everything said around them, and to keep the conversation upbeat and hopeful. It really makes a difference.

In 2005 I came across an article in the *New England Journal of Medicine* about communicating with a comatose patient using a special kind of MRI scan that measures brain activity. I was happy to see that the routine experience of ICU nurses and doctors and the families of loved ones in comas was getting the rigorous scientific attention it deserves. Awareness in even the deepest level of coma is common, and we are beginning to take it seriously and study it with the tools of modern neuroscience.

The Science Behind Awareness in Deep Coma

A few years ago, a team of Cambridge scientists got a chance to find out whether people in the deepest sort of coma could also communicate, given a means to do so. The answer they found has radically changed our understanding of comas. But first, a word about comas in general.

A serious brain injury might result in one of three levels of unresponsiveness:

1. *Coma* just means that the patient shows no sign of being awake or aware. People in comas have closed eyes and don't respond purposefully to external stimuli (that is, they don't talk, obey commands, or purposefully move their arms or legs). A coma typically lasts for several weeks. After that, the patient may either awaken or progress to a *persistent vegetative state* (PVS) or else a *minimally conscious state* (MCS).

    In a persistent vegetative state, the patient appears awake but shows no sign of meaningful interaction with the environment. People in PVS open their eyes and experience sleep-wake cycles. They continue to show basic reflexes; for example, they shrink from pain and startle at loud noises. But they don't show the purposeful interaction that suggests thought processes. They don't respond

to voices or follow objects with their eyes, and they show little sign of emotion.
2. When this condition persists for more than four weeks, it may be called a *continuing vegetative state* (CVS). When it persists for more than six months (typically from a non-traumatic injury to the brain such as a near drowning, hypoxia, or stroke) or more than twelve months (typically from traumatic brain injury), it is sometimes called a *permanent vegetative state.*
3. There is a third, more recent diagnostic category, the minimally conscious state. Patients show some inconsistent awareness. Sometimes they can respond meaningfully in simple ways to the environment, such as moving a finger when asked to do so. Often MCS follows a coma or PVS. Sometimes it is a stage of recovery from coma or PVS. Other times, unfortunately, it is not.

These altered levels of consciousness are diagnosed using a battery of tests, including bedside examination by a neurologist or neurosurgeon, CT scans, MRI scans, PET scans of the brain, and electroencephalograms (EEGs), which track brain waves.

All these altered levels of consciousness should be distinguished from *brain death*. Brain death is the biological death of all parts of the brain above its junction with the spinal cord. When brain death occurs, a patient is considered clinically dead even though the heart may still beat. All these alterations of consciousness must also be distinguished from *locked-in syndrome*, where a brain injury (usually due to a small stroke in the pons, which is a part of the brain stem) results in complete paralysis except for the ability to partially move the eyes. A locked-in patient can appear comatose or in PVS but is completely awake and simply unable to interact.

The Mind Is Hard to Just Put Out

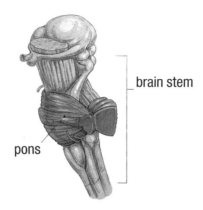

Neurologists acknowledge that terms like "vegetative," to describe a person in a coma, sound dehumanizing. Based on what we know today, they are searching for better ones. *Unresponsive wakefulness syndrome* is perhaps a better way of describing the condition.[1] We shall shortly see why "wakefulness" is a key word in the proposed new term.

Getting Away from the Dehumanizing Assumptions

The concept of persistent vegetative state—that a human being with severe brain damage could be a "shell" without an internal mental state—was formulated in the 1970s. It was officially recognized as a diagnosis in the 1990s. Since then, however, several remarkable research studies have challenged the certainty that patients in PVS lack internal mental states.

In September 2006, Cambridge University neuroscientist Adrian Owen and his colleagues published a landmark paper titled "Detecting Awareness in the Vegetative State."[2] The authors studied PVS sufferers whose brain injuries were so massive that there was no external evidence of internal mental functions. Their PVS was even deeper than a deep coma.

People in comas appear to be asleep but cannot be awakened.

People in PVS, on the other hand, can show stereotypical reflex responses (blinking, grimacing, withdrawing an arm from pain). But they have such severe brain damage that they have been assumed by doctors to have no mind at all—just a collection of reflexes.

These patients in PVS appear awake—they lie in bed with their eyes open—but they don't laugh, cry, talk, grasp objects, or do any other things that we ordinarily associate with an internal mental state. Thus the "physical shell without a mind" dictum, conventional wisdom in medicine since the 1990s, sounded convincing. And the dictum has had consequences too. In the United States in 2005, a young woman in PVS, Terri Schiavo[3] (1963–2005), was deliberately starved to death because her husband and some of her doctors believed that she no longer had any mental state at all.

Unlocking the Door to the Mind

Owen and his team set out to determine whether it is true that the mind truly cannot function in PVS. It wouldn't be easy, because people who have massive brain damage are usually also profoundly *physically* handicapped—largely paralyzed, often blind, perhaps deaf. If mental states exist, how could they be recognized? Dr. Owen decided to use functional magnetic resonance imaging (fMRI) of the brain to search for fluttering evidence of a mind.

fMRI doesn't measure neuron activity; it measures changes in blood flow in brain regions that correspond to some extent to the activity of the region's nerve cells. By visualizing blood flow in real time, researchers study the changes that correspond to brain activity in both humans and animals in a variety of situations. Thinking about things makes the fMRI "light up" in recognizable patterns. Despite the complexity and subtleties of interpretation, fMRI research has proven a valuable research and clinical tool.

Owen and his colleagues tested a thirty-four-year-old woman who had suffered massive brain injury in a car accident a year earlier. After a battery of standard tests, including MRI scans, CT scans, EEGs, and detailed bedside examination, she was diagnosed with PVS, showing no evidence of any mental state at all. She was, experts declared, a human body without a mind.

But Owen tried a different approach. He fitted her with a headset and put her in an MRI machine. While the machine was imaging her brain activity, he asked her to think about various things: "Imagine playing tennis," or "Imagine walking across a room."

Her severely damaged brain showed areas that lit up. But what did that mean?

To find out, Owen asked volunteers with normally functioning brains to undergo the same test—the headsets, the fMRI machine, and the questions. Remarkably, the normal volunteers *had exactly the same pattern of brain activation* as the young woman in PVS. So a woman in the deepest possible coma showed physical signs of awareness of his words that was indistinguishable from that of the mentally normal volunteers.

A careful scientist, Owen added another step to the experiment. He put her back in the fMRI machine and used the same words to ask her to imagine playing tennis or walking across a room. Except this time he scrambled the words so that they made no sense—"tennis imagine playing." He did the same test with the neurologically normal volunteers. In both cases, scrambling the words caused the fMRI brain activation patterns to cease. Clearly, the young woman in PVS *understood* the questions she was being asked, as did the volunteers.[4]

Follow-up research on other patients confirmed the general pattern. In 2010, Owen, Steven Laureys of the University of Liège, and other researchers published an article in the *New England Journal of Medicine*, again reporting on the use of fMRI to evaluate awareness in patients diagnosed in PVS.[5] Of fifty-four such patients, five could respond (as demonstrated by fMRI imaging) to requests to think about certain scenarios. One was even able to answer yes-or-no questions accurately (again via fMRI imaging).[6]

The one who could answer questions was a twenty-nine-year-old man, brain-damaged in a car crash in 2003. He had been in a coma for two years before slipping into PVS. Apart from an occasional blink, he seemed unaware. Owen and Laureys's team wanted to question him using fMRI, but the brain responses that mean "yes" or "no" are both complex and hard to distinguish. So instead, the inventive researchers asked him to think about playing tennis if he meant yes and moving around at home if he meant no. Thoughts of tennis stimulate "spatial" areas high in the brain, but thoughts of moving around the house spark "motion" areas at the brain's base. They then asked him six simple questions about his life, including his father's name and whether he had sisters.

The scans picked up his thoughts within five minutes, *and he got every question right*. Owen told media, "We were astonished when we saw the results of the patient's scan and that he was able to correctly answer the questions that were asked by simply changing his thoughts."[7] Similarly, in 2011, researchers publishing in *The Lancet* showed that, for three of sixteen patients diagnosed in PVS, a bedside EEG could demonstrate awareness.[8]

In 2022, *Scientific American* reported the story of thirty-year-old Maria Mazurkevich, who showed no sign of consciousness in the hospital after a blood vessel in her brain ruptured. But her family insisted she was still "in there." Could that be tested?

The medical team gave her an EEG—placing sensors on her head to monitor her brain's electrical activity—while they asked her to "keep opening and closing your right hand." Then they asked her to "stop opening and closing your right hand." Even though her hands themselves didn't move, her brain's activity patterns differed between the two commands. These brain reactions clearly indicated that she was aware of the requests and that those requests were different.[9]

Fortunately, she recovered and was able, after a year, to go to work as a pharmacy assistant. Some researchers refer to the state she had experienced as *covert consciousness*. It's not known how many comatose people are trapped in covert consciousness, but a 2019 estimate suggests 15 to 20 percent.[10] A 2024 paper in the *New England Journal of Medicine* put the number of covertly conscious comatose patients at 25 percent.[11]

Motivated in part by these sobering discoveries, medical researchers have begun to define the minimally conscious state (MCS)—the new diagnostic category for comatose people who show evidence of thought processes[12]—in more detail. There may be between 112,000 and 280,000 patients with MCS in the United States alone.[13] They are aware of their environment to some extent and have some meaningful capacity to respond to it, despite having largely nonfunctional brains.

How Coma Helps Us Understand the Mind Better

As we have seen, many neuroscientists believe that the mind is simply the physical activities of the brain. Beyond that, the mind does not really exist. Their position is often described as the *materialist* view. Yale University neurologist Steven Novella offers five propositions in support of the materialist view:

> *If the mind is completely a product of the material function of the brain, then:*

1. There will be no mental phenomena without brain function.
2. As brain function is altered, the mind will be altered.
3. If the brain is damaged, then mental function will be damaged.
4. Brain development will correlate with mental development.
5. We will be able to correlate brain activity with mental activity—no matter how we choose to look at it.[14]

Now, here are five alternative propositions. We start with the view that the mind is not simply the physical activities of the brain, that it also has an independent existence. This is often called the *dualist* view.

> *If the mind is partly the product of the material function of the brain and partly the product of something that is beyond nature, then:*

1. There will be some mental phenomena without brain function.
2. As brain function is altered, the mind will not necessarily be altered.
3. If the brain is damaged, then mental function will not necessarily be damaged.

4. Brain development will not necessarily correlate with mental development.
5. We will not always be able to correlate brain activity with mental activity—no matter how we choose to look at it.

Now let's look at it in relation to coma. Owen, Laureys, and others have found relatively sophisticated internal mental states (for example, the ability to imagine scenarios and to provide yes-or-no answers to questions) in the presence of severe brain damage. Here we are talking about damage that is so severe that highly skilled neurologists and neurosurgeons had inferred by bedside examination alone that the patient *had no mind at all*. That should be deeply perplexing from a materialist perspective. After all, PVS is the most severe form of brain damage short of brain death.

If materialism is correct, how can a person who has suffered such severe brain damage as to be unable to respond meaningfully to the environment provide subtle and meaningful mental responses, when pressed? Keep in mind that materialism, as set out by Novella, is a totalistic argument. Even one exception disproves it. We dualists can readily account for such mental function: The mind generally depends on the function of the brain. But it can also, at times, function independently.

Materialist neuroscientists have sometimes responded by pointing out that the fMRI evidence for awareness is itself a material phenomenon—thinking activates changes in the blood flow in the brain.[15] That's true, but it misses the point. The only way we can find out what is going on in other people's minds under any circumstances is some sort of behavior or response on their part. A response where they do something we ask them to do is especially informative. The response in these cases, as imaged by fMRI, was *not* predicted by materialist philosophy. On the contrary, it seems that the mind, in addition to being a unity, can function at a surprisingly high level (for example, to imagine complex activities and to answer questions) despite the most severe form of brain damage short of brain death.

Consider Owen and Laureys's PVS patient mentioned earlier, who used thinking about playing tennis if he meant yes and moving around at home if he meant no. This means that not only was the man able to answer yes-or-no questions while in the deepest level of coma, but he could also answer in a very clever way—he understood, remembered, and consistently used a complex code (that is, thinking of different activities) to express himself.

But now what if the brain is actually dying? Can the mind function then?

A Sudden Light as Death Approaches

When a loved one journeys toward death and bodily resources fail, the lights may seem to be going out, one by one. But then...they may come back on again, briefly. Journalist Alex Godfrey described his grandmother's death in the *Guardian* in 2021:

> For a week she was barely conscious, but on the Sunday when my parents, cousin and I visited, she was sitting up in bed, smiling as we walked in. For the next two hours she laughed and joked, completely cognitive, coherent...lucid. A lifetime of memory had returned, and we took advantage of it as she regaled us with escapades from her past. My mum, who knew many of them, quietly verified them. Her funny, eloquent, vibrant mother had returned. "It all came back to her in one rush," remembers my mum. "It was like a bolt of lightning. The clouds cleared." After we left that afternoon, my grandma slipped back into a semi-conscious

state, soon not knowing who my mother was, and died within days.[16]

Such accounts are not uncommon. Science writer Jordan Kinard reported in *Scientific American* in 2023 that for decades, researchers, hospice workers, and family members have watched "with awe" as victims of dementia suddenly find their memories and personalities again, just before they die.[17]

Of course, historical and traditional accounts of such deathbed scenes abound. But in recent years, medical researchers have also started studying this sudden, remarkable lucidity—*terminal lucidity*[18]—weeks, days, or hours before death.[19] Is it, as a materialist might say, mere noise from a dying brain? Or is it a signal, intimating what lies beyond? One research team confirms, "It happens unexpectedly: a person long thought lost to the ravages of dementia, unable to recall the events of their lives or even recognize those closest to them, will suddenly wake up and exhibit surprisingly normal behavior, only to pass away shortly thereafter."[20]

Skeptics Have Adopted a Wait-and-See Approach

Even people who might prefer to believe that terminal lucidity is just random brain noise admit that they are not sure. Science writer Jesse Bering tells us, "I'm as sworn to radical rationalism as the next neo-Darwinian materialist. That said, over the years I've had to 'quarantine,' for lack of a better word, a few anomalous personal experiences that have stubbornly defied my own logical understanding of them."[21] Similarly, at *Discover* magazine, "Neuroskeptic" offers readers a remarkable account of terminal lucidity from the early twentieth century, stating, "I do not believe in miracles and this story didn't change my mind on that score. However, unless we reject the whole story as a fiction, it is surely one of those 'anomalies that neuroscience ought to be able to account for.'"[22]

Well, yes. Neuroscience ought to be able to account for terminal lucidity. But why should that mean reassuring the world that someday we will prove that it is just random brain noise? That's the trouble with materialism as a foundation for neuroscience. It comes to mean an endless search for materialist explanations that don't really fit the evidence instead of seeing what we can learn from the evidence.

Just what is happening in the human brain during bouts of terminal lucidity remains unclear.[23] Sam Parnia is director of research into resuscitation after heart attacks at New York University's Langone Medical Center. He notes that in one multicenter study, one in five survivors of heart stoppage reported a lucid experience.[24] He suggests that the lucidity is triggered by a spike in brain activity due to the loss of oxygen. But he does not dismiss these experiences as mere noise: Rather, in his view, the dying process "gives you access to parts of your brain that you normally can't access."[25] As he told science media, the study—of which he is one of the authors—"found these experiences of death to be different from hallucinations, delusions, illusions, dreams, or CPR-induced consciousness."[26]

Certainly, what's happening seems to go well beyond a mere spike in brain activity. George Mason University researcher Andrew Peterson, who studies terminal lucidity,[27] enlarged on that point for *Scientific American*: "There seems to be clear evidence that they're aware not merely of their surroundings...but additionally understanding what their relationships to other people are—be it the use of a nickname or a reference to a longstanding inside joke."[28]

Sometimes lucidity becomes a way of saying goodbye. Australian palliative care doctor Will Cairns notes that one of his patients, dying at home, was unresponsive for two days while his son was traveling to see him. But when the son arrived, he woke up to talk with him for several hours. Then he became unresponsive again, and a few hours later he died.[29] Cairns asks:

> How many times have nurses told us at morning handover that early the previous evening one of our patients who seemed stable and had been sleeping peacefully most of the time in our palliative care centre had roused themselves to an alertness not seen for ages and asked the nurses to summon their family for a meeting? After a period of conversation, the patient has gone back to sleep, and died later that night.[30]

Modern palliative care furnishes many accounts that sound like the vast traditional literature on last words—a final communication at the point of death.[31] The mind, sensing that the body is failing, rallies briefly for a purpose. Needless to say, such lucid episodes imply that the mind is more than the disjointed activities of a failing brain.

Paradoxical Lucidity—The Mind Suddenly Surfaces When Death Is Not Imminent

Among people afflicted with severe dementia, sometimes the clouds part well before death. Lucidity can emerge as a sudden unexpected ability to communicate—often in response to an emotional trigger such as a remembered voice.[32] Because there is currently no medical explanation, researchers call it *paradoxical lucidity*.

That happened to Denyse O'Leary's very aged father, who had been diagnosed with dementia. Generally, he did not appear to notice his surroundings very much. But one day, observing his daughter running around, preoccupied with many tasks, he suddenly remarked clearly (and shockingly), "Denyse, when I am gone, you won't have to work so hard anymore"—an obvious reference to his looming mortality and his modest estate. But then he nodded off.

In one study of 124 people with dementia, more than 80 percent[33] experienced a return of lucidity at some point, "complete remission with return of memory, orientation, and responsive verbal ability," as reported by observers.

According to one research team's records, more than 90 percent of people with severe dementia who showed terminal lucidity died within seven days, 41 percent within one or two days, and 15 percent within two hours.[34] But the team also acknowledged that there is a significant gap in the research. John O'Leary, for example, lived for another two years, dying at ninety-nine.

The trouble is, a focus on lucidity only when it occurs near death may underrepresent its true prevalence. Friends and relatives tend to gather once they know that a loved one who suffers from dementia is dying. An episode of lucidity is likely to be observed during a round-the-clock death watch. But if an unvisited loved one had dementia for years, a number of episodes of *non*-terminal lucidity might have occurred but never been noticed or recorded.[35]

Ignoring lucid episodes that don't presage imminent death may come at a cost. Alzheimer's disease, the leading cause of dementia, affects roughly 6.5 million people in the United States (2023).[36] If episodes that are not associated with imminent death were identified and tracked more often, they might provide new avenues for dementia research. Right now, all we know about sudden bouts of lucidity is that they happen. More research would be required to determine if some elements of naturally occurring remissions can be incorporated into therapy.[37]

Stephen G. Post, director of the Center for Medical Humanities, Compassionate Care, and Bioethics at the Stony Brook University School of Medicine, has been an advocate for "deeply forgetful people," as he prefers to call those with profound dementia, since the early 1990s. For example, he was a pioneer in the use of therapy dogs in dementia support programs. When discussing these sudden awakenings, he points out a little-discussed fact: We know very little about human memory overall. "Nobody really knows what biographical memory is, physically or metaphysically, and so it remains a mystery."[38] To take just one example, it's not clear how our massive visual memory can even be stored physically in the human brain.[39] If we don't know very much about memory, we are not well placed to understand forgetfulness.

In any event, the relationship between memory and the brain is much more plastic than we might think. Pediatric neurologist John Lorber (1915–1996) reported in the 1970s that some adults who suffered hydrocephalus (water on the brain) as children had good memories as adults, though they retained no more than 5 percent of normal brain tissue. His research was dismissed[40] in *Science* in 1980 as unscientific and "overdramatic." But in the decades since, good memory without much brain tissue has been independently confirmed.[41] That topic, like paradoxical lucidity, has received surprisingly little study[42] when we consider the implications in terms of care for our brain-injured loved ones.

What Coma and Sudden Lucidity Tell Us About the Human Soul

Researchers have learned that the mind can, at times, break through massive brain damage or disease, especially near death. And that sometimes the result is a final burst of intense thought. In his book *Dignity for Deeply Forgetful People* (2022), Post recounts receiving an email from a fellow panelist concerned with dementia issues, who told him that the panel discussion "reminded me of a moment with my beloved mother, a poet, author, and something of a philosopher. In that late stage when words are gone except for those very occasional moments, she looked at me intently and said forcefully, 'God, physics and the cosmos.'"[43]

But if the diseased brain is all there is to human consciousness, why should demented people so often become so lucid, especially just before death? Terminal/paradoxical lucidity is a direct challenge to materialist theory.

Meanwhile, we have now added a bit to our picture of the relationship between the mind and the brain. The mind can sometimes communicate when the brain is all but dead. It can burst, both suddenly and fully, from a ravaged and dying brain. But now, let's offer the mind another challenge: What if two minds must share a body or large parts of a body? Yes, it happens... Will they merge?

CHAPTER 4

When Two Minds Must Share Body Parts

WHEN TWINS ARE BORN conjoined but inseparable, they sometimes share parts of their brains. How does that affect their individuality, their minds, their souls? What happens in these instances is remarkable and revealing.

Krista and Tatiana Hogan, born in 2006 in British Columbia, Canada, are joined at the head (craniopagus). They cannot be surgically separated because they share critical areas deep in the brain. Pediatric neurologist Juliette Hukin told media in 2012 that they have a connection in the brain, "a bridge of white matter that connects the mid brain and the thalamus in both twins" (a unique *thalamic bridge*).[1]

The thalamus is a critical part of the brain that mediates wakefulness and motor and sensory function. Thus, sharing one gives the girls the ability to sense and even move large parts of each other's bodies. A

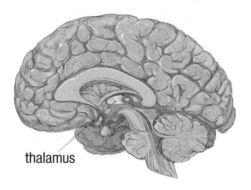

thalamus

Canadian Broadcasting Corporation documentary spells out what that means:

> Krista and Tatiana Hogan share the senses of touch and taste and even control one another's limbs. Tatiana can see out of both of Krista's eyes, while Krista can only see out of one of Tatiana's. Tatiana controls three arms and a leg, while Krista controls three legs and an arm. They can also switch to self-control of their limbs. The twins say they know one another's thoughts without having to speak. "Talking in our heads" is how they describe it.[2]

Specifically, Krista has motor control and sensation in both of her own legs *and of Tatiana's left leg*.

The documentary shows Felicia, their mother, lightly touching Tatiana on different parts of her body while Krista's eyes are closed. Felicia asks Krista, who keeps her eyes shut, where her fingers are making contact. "Her cheek," Krista responds correctly. Then, "Her knee."[3] Krista seems to be in no doubt about the fact that it is *her sister's* knee, even though she can feel the sensation in it herself.

When Two Minds Must Share Body Parts

Tatiana likewise has motor control and sensation in both of her own legs and arms and has motor control and receives sensation from *Krista's right arm*.

Are the twins really only one person, then? Some philosophers are prepared to consider that possibility:

> When twins share significant amounts of brain tissue, like Tatjana and Krista Hogan...who can see through each other's eyes and feel each others' pin pricks, it may be unclear whether there is one partially split mind or two separate minds and, thus, one or two of us, according to psychological theories. On biological theories, however, in some of these cases, there is definitely only one of us with a more or less split mind.[4]

But Sam Kean, author of *The Tale of the Dueling Neurosurgeons* (2014), disagrees: "Even though, on some physiological level, they're experiencing the same thing, their response to it is quite different," he says. "It is, I think, evidence that they are individual people, despite sharing parts of their brain."[5] He notes, for example, that the sisters don't share preferences. Krista loves ketchup; Tatiana hates it. And one really likes corn, though it causes the other to break out in hives.[6]

What They Don't Share

Krista and Tatiana Hogan share sensations, but despite the anatomical connection of their brains, they do *not* seem to share abstract reasoning. Notably, there is no report that they share concepts such as the arithmetic or logic that they learn in school, as if their minds were the same system. There's no indication that they can split studying—have Tatiana study geometry and Krista study calculus simultaneously—and have both twins simultaneously learn both subjects. They also have obviously separate wills—they disagree about quite a bit. That is likely due to the immaterial nature of abstract thought. Wilder Penfield would probably have predicted that, based on his research on awake brain patients.

Inseparability is a reality for the twins, but like "youth" or "career," it is also an abstraction. The twins had to learn its meaning the hard way when they were quite young. As the CBC feature recounts:

> When they were little, they used to try to pull their heads apart. Their mother always told them they were stuck, so they would have to work things out. But as they've gotten older and the frustrations mount, they still fight. As they freely admit, some days they don't like being together. "She's annoying," says Tatiana, who promptly gives her twin a reassuring hug.[7]

So each twin had to separately realize the concept of *inseparable* for herself and internalize what it meant for her own future. ("Future" is another abstraction, of course.)

What Conjoined Twins Can Tell Us About the Soul

A traditional picture of the soul sheds light on the twins' experience. Thirteenth-century theologian Thomas Aquinas[8] (c. 1225–1274) adapted the approach of the ancient philosopher Aristotle (384–322 BC) to the Christian tradition in this way:

The human mind consists of three types of powers: vegetative, sensory, and rational. *Vegetative* powers include control of body functions that keep us alive. We are rarely conscious of these basic powers, which we share with plants. We also have *sensory* and *motor* powers, the ones Penfield could stimulate with his electrodes (sensation, motion,

When Two Minds Must Share Body Parts

memory, emotion, and so on). We share them with other animals and we are very conscious of them.

But human beings have *rational* powers as well, which are not shared by plants or animals. These are, as we have seen, the ability to think abstractly and choose ethically. With conjoined twinning, as with split or largely absent brain parts, abstract reasoning and ethical choice can be comparatively unaffected by very unusual brain arrangements.

Both vegetative and sensory powers are *material*; they originate in physical processes and are tightly linked to brain function. But rational powers like abstract thought and moral choice are *immaterial*, even though their usual expression requires matter. Thus the human mind or soul has both material and immaterial parts. When we talk about the human *mind* or *soul*, we mean both parts together, though we usually focus on the immaterial part.

The abilities that Tatiana and Krista share are *material powers* of the brain—for example, their abilities to feel and move each other's limbs and see with each other's eyes. We should expect them to share these parts because they share brain matter. What they don't seem to share is the immaterial aspects of the soul—abstract reasoning and personal identity, individuality, free will, and so on. They are separate souls who share some brain tissue.

From a description in a Canadian magazine:

> There are times [Tatiana] wants to follow her own interests, and fights begin. "Krista will want to go to sleep, and Tatiana wants to watch TV," says their mother. "And then [Krista] gets mad. That's when most of the war wounds happen."
>
> Mostly, out of necessity, they work together in seamless fashion. One of their favourite jobs is unloading the

dishwasher. From one pair of hands to the other, to whoever is putting the dishes up in the cupboards.[9]

In fact, it is their obvious distinctness that is the basis for our amazement at the perceptions and motor control that they do share.

Conjoined Twins Typically Share These Qualities

While much publicized, inseparable conjoined twins are statistically rare—between one in fifty thousand births and one in two hundred thousand births. Most do not survive birth or not for long. Thus, they are not widely studied.[10] Most of those who do survive birth are female (three out of every four),[11] but of the 40 to 60 percent who are stillborn, most are male.[12] In other words, conjoined twinning is not more common among girls; rather, most survivors are female. While each pair may have a different life situation, as we shall see, they all remain individual souls.

Carmen and Lupita Andrade, for example, born in 2002, are joined at the torso, sharing a pelvis, liver, bloodstream, and reproductive system. Each has two arms, but only one leg (Carmen controls the right leg; Lupita, the left). They live in Connecticut and have taken training as veterinary workers.

They say they can feel each other's emotions. Lupita told an interviewer, "I can feel when Carmen is anxious or about to cry. It's that same stomach drop." But they don't share all emotions. Carmen has a boyfriend, but Lupita, who professes no interest in sex, likes Carmen's boyfriend simply as a person.

About their overall situation, Carmen told media in 2023:

> Sometimes at the end of the day, we're just exhausted and we don't want to talk. That's when we'll go on

different devices and do our own thing. I have my laptop to do schoolwork, and Lupita will put on headphones and listen to music or go on her phone. We've been conjoined our whole life, so it's not like we miss our independence. It's all we've ever known, right?[13]

They come off as independent siblings who form separate abstract ideas, living with a conjoinedness that includes physical awareness of each other's emotions. That makes sense if emotions are among the *material* powers of the brain.

Perhaps the best-known conjoined twins in the United States are fifth-grade math teachers Abby and Brittany Hensel, born in New Germany, Minnesota, in 1990. Sharing a single body, they must coordinate everything they do. According to *Psychology Today*:

> Each has a separate head, heart, lungs, spine, stomach, and spinal cord, but they share two arms, legs, large intestine, bladder and reproductive organs. Given that they share a body, and most importantly, a single pair of arms and legs, they have to coordinate everything they do. In fact, each twin manages only one side of their body, making all movements an amazing act of teamwork, yet they can walk, run, swim, play basketball, and even drive a car.[14]

For example, as teachers, they share an email account. With each twin controlling one hand, they can type seamlessly in sync.[15]

Now in their midthirties, they've negotiated many "singular or plural?" issues. For example, they passed driving tests separately and each holds her own license.[16] But they only receive one salary at Sunnyside Elementary in New Brighton, Minnesota, because they are said to be "doing the job of one person." In fact, the staff directory, accessed in 2023, shows their two faces but describes them as (one) "teacher."[17] And they aren't happy with that.

Abby Hensel has offered, "As maybe experience comes in we'd like to negotiate a little bit, considering we have two degrees and because we are able to give two different perspectives or teach in two different ways."[18] For one thing, they can hold separate conversations, a trait quite useful for answering student questions.[19] And yet, not surprisingly, they also show a common "twin" trait; from a lifetime of experience, they can finish each other's sentences.[20] Through it all, they remain two separate minds sharing a body.

Conjoined Twins Can Lead Very Different Inner Lives

As of 2024, the oldest living conjoined twins are Lori and Dori/George Schappell, born in Sinking Spring, Pennsylvania, in 1961.[21] Joined inseparably at the head like Tatiana and Krista Hogan, they are said to share 30 percent of their frontal lobe brain tissue. But they are clearly quite different. Lori is sexually active[22] with boyfriends, but Dori, who suffers from spina bifida, chose to identify as a male (George) as of 2006. Before that, Dori had pursued a career as Reba, a country-and-western singer,[23] with Lori's help. Lori also worked in a hospital laundry[24] while her sibling read books. When they emphasize to interviewers that they are very different people, there is little reason to doubt it.

No two conjoined twin situations are alike, but maintaining individuality as human beings does not appear to be the challenge we might have expected. That makes sense if the individual mind is a natural unity; it remains a unity even when sharing parts of a physical body with another mind.

But now, a bigger challenge: Can the human mind ever act completely independently of the brain? In chapter 3, we saw that people in a persistent vegetative state, which is just short of clinical brain death in terms of consciousness, have been able to communicate with researchers. But what if the brain *is* clinically dead? Can the mind function then? That's the basic claim for verified near-death experiences.

CHAPTER 5

The Human Mind Beyond Death

IN 1991, THIRTY-FIVE-YEAR-OLD PAM Reynolds, an American singer-songwriter from Atlanta, began to feel dizzy and have trouble speaking. She also had difficulty moving her arms and legs. When she sought medical help, a CT scan and an angiogram showed the dismaying cause of her symptoms: A bulge in the wall of a blood vessel the size of a golf ball rested silently and ominously at the top of her basilar artery.

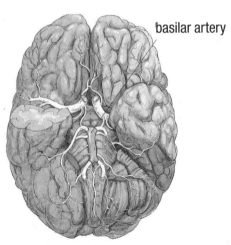

The Human Mind Beyond Death

The basilar artery is the main artery supplying blood to the brain stem—the most vital part of the human brain. Not only was the bulge (an *aneurysm*) ballooning at the top of the brain's blood supply, but it was also in danger of bursting. If it burst, she would either die or suffer catastrophic brain damage. But conventional neurosurgery might be too risky.

Reynolds was referred to the world's leading aneurysm neurosurgeon, Dr. Robert Spetzler, director of the Barrow Neurological Institute in Phoenix. Spetzler judged that her only hope was an extraordinary technique called hypothermic cardiac arrest (a "standstill procedure"). She would be placed under general anesthesia and her body would be chilled to 60°F (15.6°C). Then her heart would be stopped so that no blood flowed to her brain. And her brain would be *drained of blood*.[1]

In other words, Spetzler's strategy was literally to create brain death due to a low body temperature and no blood circulation. That would give him a thirty-minute window to get rid of the aneurysm while her brain was like an object, unable to react. She would be protected from permanent brain damage both by her hypothermia and by the medications administered during the time she was brain dead. Later, her body would be warmed and her heart restarted, so blood flow to her brain would resume.

She consented. So one day in August 1991, she was taken to the operating room at 7:15 a.m. and given several intravenous catheters and general anesthesia. After she was under anesthesia, the team lifted her body onto the operating table and tied it down securely. Her eyes were lubricated to prevent desiccation and then sealed shut. A breathing tube was passed through her mouth into her windpipe to provide her with oxygen. Doctors and nurses in the operating room then took an hour and twenty-five minutes to connect instruments to her body to monitor all of her vital signs. That meant monitoring her brain waves and signs and the dramatic physiological changes that the operation required.

Doctors attached wires and catheters to monitor her heartbeat, blood pressure, blood oxygen, and body temperature. That way they could safely cool her body and protect her from the effects of stopping her heart for the duration of the operation. EEG electrodes were then taped to her scalp to continuously monitor her brain waves and confirm when her brain was clinically dead.

To further confirm that her brain would be dead, molded earplugs were inserted into both of her ears. They emitted 100 dB clicks every second that would continuously stimulate her hearing nerves and thus be recorded in her brain waves. The sound of the clicks was about as intense as the noise generated by a loud hair dryer on the highest setting, blowing into each ear. When she stopped responding to these loud clicks, medical personnel would know that all of her brain activity had ceased. The continuous loud clicks also meant that even if she were conscious, she could not hear conversations in the operating room.

Then her skull was clamped into a steel pinhead holder and her body was turned on her left side to position her properly for the surgery. Except for the side of her skull and her groin, her entire body was covered with thick surgical drapes. More than twenty doctors, nurses, and technicians worked together as the surgery began. Dr. Spetzler incised Reynolds's scalp and used a Midas Rex surgical drill to carve a window into her skull to expose her brain.

It was at this point that Pam Reynolds's near-death experience began. She later recounted:

> The next thing I recall was the sound: it was a natural D. As I listened to the sound, I felt it was pulling me out of the top of my head. The further out of my body I got, the more clear the tone became. I had the

impression it was like a road, a frequency that you go on... I remember seeing several things in the operating room when I was looking down. It was the most aware that I think that I have ever been in my entire life... I was metaphorically sitting on Dr. Spetzler's shoulder. It was not like normal vision. It was brighter and more focused and clearer than normal vision... There was so much in the operating room that I didn't recognize, and so many people.

She noticed that her head had not been entirely shaved, as she had expected. The only part that was shaved was the planned incision area.

The saw thing that I hated the sound of looked like an electric toothbrush and it had a dent in it, a groove at the top where the saw appeared to go into the handle, but it didn't... And the saw had interchangeable blades, too, but these blades were in what looked like a socket wrench case... I heard the saw crank up, I didn't see them use it on my head, but I think I heard it being used on something. It was humming at a relatively high pitch and then all of a sudden it went *Brrnrrrrrrrr!* like that.[2]

At this point, Dr. Spetzler had opened the dural covering over her brain and was using the operating microscope to carefully dissect along the surface of her brain deep down to the top of the basilar artery in front of her brain stem. While he was exploring the aneurysm, a female cardiac surgeon exposed the artery and vein in Reynolds's right groin. These blood vessels proved too small to accommodate the catheters that would be needed when they stopped her heart (the cardiac standstill).

The cardiac surgeon then exposed her left groin arteries, which turned out to be big enough. Reynolds recalled hearing the matter discussed:

> Someone said something about my veins and arteries being very small. I believe it was a female voice and that it was Dr. Murray, but I'm not sure. She was the cardiologist [*sic*]. I remember thinking that I should have told her about that.[3]

But remember, because of the loud clicks in her ears, she could not physically have heard anything.

When Dr. Spetzler examined the aneurysm under the operating microscope, he noted (in his dictated operative notes) that it was "extremely large and extended up into the brain." As the team had expected, it could not be repaired without cooling Reynolds's body. That meant stopping her heart and draining the blood from her brain.

At 10:50 a.m. the cardiac surgery team began cooling Reynolds's blood using the heart–lung machine. In ten minutes Reynolds's body temperature had dropped 25°F (14°C). Within fifteen minutes (after a dose of potassium chloride), her heart stopped. Her brain waves, which were measured continuously with electrodes on her scalp, ceased.

The loud clicks in her ears no longer caused any electrical response in her brain stem as her body temperature plummeted to 60°F (15.6°C). Dr. Spetzler tilted the operating table so that Reynolds's head was above her feet and drained the blood out of her brain.

So at 11:25 a.m. Pam Reynolds was, for all intents and purposes, brain dead. There is no natural way that she should have been conscious of anything at all. Yet she recalled later:

> There was a sensation like being pulled, but not against your will. I was going on my own accord because I

wanted to go. I have different metaphors to try to explain this. It was like the Wizard of Oz—being taken up into a tornado vortex, only you're not spinning around like you got vertigo. You're very focused and you have a place to go. The feeling was like going up in an elevator real fast. And there was a sensation, but it wasn't a bodily, physical sensation. It was like a tunnel but it wasn't a tunnel.

She became aware of her deceased grandmother calling for her, but the sound was clearer than natural hearing provides. She continued down the tunnel until she came across a tiny, very bright spot of expanding light:

I noticed that as I began to discern different figures in the light—and they were all covered with light—they *were* light, and had light permeating all around them—they began to form shapes I could recognize and understand. I could see that one of them was my grandmother. I don't know if it was reality or projection, but I would know my grandmother, the sound of her, anytime, anywhere.

Everyone I saw, looking back on it, fit perfectly into my understanding of what that person looked like at their best during their lives.

It turned out that they had a message for her:

They would not permit me to go further... It was communicated to me—that's the best way I know how to say it, because they didn't speak like I'm speaking—that if I went all the way into the light something would happen to me physically. They would be unable to put this me back into the body me, like I had gone too far and they couldn't reconnect. So they wouldn't let me go anywhere or do anything.[4]

In truth, she wanted to go back, to continue raising her children.

Meanwhile, back in the operating room: Deprived of blood flowing to her brain, the aneurysm at the top of Reynolds's basilar artery naturally deflated. That gave Spetzler an opportunity to place an aneurysm clip along its base and occlude it. He then carefully dissected the empty aneurysm sac, removing it from her brain. That relieved the compression of the brain stem that was causing her symptoms. The cardiac doctors then began to circulate and warm her blood and thus to bring her slowly back to life.

As her body temperature rose and her blood began to circulate again, the electrical signals in her brain returned. Thus her brain stem showed electrical activity in response to the loud clicks in her ears. In the meantime, Reynolds experienced a sense that her deceased relatives were feeding her in some way: "I definitely recall the sensation of being nurtured and being fed and being made strong. I know it sounds funny, because obviously it wasn't a physical thing, but inside the experience I felt physically strong, ready for whatever."[5]

At noon, as her body warmed, her heart began to fibrillate. This was not a good thing; it was a serious problem. If it could not be corrected, her heart could not be restarted. So the cardiac team placed electrical paddles on her chest and shocked her heart. After two jolts, her heartbeat returned. This return seems to correlate with an event in Reynolds's near-death experience: "But then I got to the end of it and saw the thing, my body. I didn't want to get into it... It looked terrible, like a train wreck. It looked like what it was: dead. I believe it was covered. It scared me and I didn't want to look at it... The body was pulling and the tunnel was pushing... It was like diving into a pool of ice water... it hurt!⁶

As Reynolds's body warmed to near normal temperature, the cardiac team turned off the heart–lung machine, catheters were removed, and the surgical team closed her wounds. At that point, Reynolds recalls, the team was playing rock music in the background, including "Hotel California," which features the line "You can check out any time you like, but you can never leave." It was true; they *were* playing that. She said later that, from her perspective at the time, it seemed "incredibly insensitive."⁷ Of course, they never imagined that *she* could hear it.

By 2:10 p.m., Dr. Spetzler noted in his surgical report that Reynolds was taken to the recovery room, intubated but in stable condition.

The Mind at Death

Pam Reynolds, who died in 2010 of heart failure at the age of fifty-three, has received a good deal of attention in the study of near-death experiences (NDEs). Unlike the vast majority of NDEs, in which absence of brain function is inferred based on routine data collection, Reynolds's case was *measured and documented* by highly sensitive instruments (EEG and brain stem auditory evoked responses). Indeed, brain death was an essential feature for the success of the operation.

During her cardiac standstill, Reynolds was "brain dead" by all three of the clinical criteria ordinarily used to diagnose the state: Her EEG was silent, her brain stem responses were absent, and no blood flowed through her brain—in fact, all the blood was drained out of her brain. Yet her NDE entailed detailed knowledge of events in the operating room during clinical brain death, details that were confirmed afterward. One young neurosurgeon who witnessed the operation and saw Reynolds afterward was so startled by her recall of the procedure that, in his own words, he "put the kybosh" on the discussion.[8]

Most NDEs occur in situations that are hardly ideal research environments. Typically, they are unplanned, chaotic situations in which a patient dies unexpectedly and a medical team struggles with emergency resuscitation. If the team succeeds, the patient may report an NDE later. But Pam Reynolds's "brain death" was deliberately planned and carried out under meticulously documented circumstances. It entailed knowledge, verified later, that she could only have had if her soul—the immortal part of the mind—functioned while her brain was dead. It is clear evidence that the mind can function quite independently of the brain.

After the publicity that Reynolds's remarkable experience generated, Dr. Spetzler told CBS's *48 Hours*, "If you would examine the patient from a clinical perspective during that hour, that patient by all definition would be dead. At this point there is no brain activity, no blood going through the brain. Nothing, nothing, nothing."[9]

On the BBC documentary examining Reynolds's NDE, *The Day I Died*, Dr. Spetzler commented:

> I don't think that the observations she made were based on what she experienced as she went into the operating theater. They were just not available to her. For example, the drill and so on, those things are all covered up. They are invisible; they were inside their packages. You really don't begin to open until the patient is completely asleep, so that you maintain a sterile environment... At that stage in the operation nobody can observe, here in that state. And I find it inconceivable that the normal senses, such as hearing, let alone the fact that she had

The Human Mind Beyond Death

clicking modules in each ear, that there was any way for her to hear through normal auditory pathways.[10]

He did not offer an explanation.

Despite the fact that Reynolds could not have had physical brain activity and that her recollections of events in the operating room were confirmed, some critics have labeled her NDE as "anesthesia awareness."[11] We will consider the credibility of a number of such materialist explanations for NDEs in the next chapter. But first, let's look at what we are learning from the study of a large number of other reported NDE experiences.

What Happens During a Near-Death Experience?

Near-death experiences have been reported with increasing frequency in modern times, probably because advances in medicine have meant that many more patients can be brought back from brain death to tell us about them. One analysis of eleven large studies shows that the incidence of modern NDEs ranges from 18 to 47 percent of resuscitations,[12] but that higher percentage may be an outlier. University of Virginia psychiatrist Bruce Greyson, a specialist in near-death experiences, concludes conservatively that the most reliable estimates range between 9 and 18 percent.[13] But when we consider how many resuscitations from clinical death occur each year now, that probably means a lot of experiences.

What we are learning is that the NDE usually has a profound impact on the experiencer. Many experiencers report that they've lost the fear of death, feel "spiritualized," and have a much greater desire to live a life of love, mercy, and care for others.[14]

It's not a matter of personality. People of a variety of personality types experience NDEs, but the experiences themselves show certain similarities. These similarities are called the *core experience*.[15] That includes a subjective sense of being dead, a feeling of peace, painlessness, pleasantness, and so on, a sense of bodily separation, perhaps a sense of entering a dark region or tunnel, encountering a presence or hearing a voice. That may include taking stock of one's life (a life review), seeing a bright and often beautiful light or being enveloped in it, seeing beautiful colors, entering the light, and encountering visible spirits. Near-death experiencers frequently report mental conversations with a Being in the light or with deceased persons, especially relatives. Those who report undergoing a life review may also report a sense of forgiveness as well as profound understanding in a world of beauty beyond nature.

It is unusual for an NDE experiencer to have all of these experiences, but most NDErs have at least several. They often enjoy dramatically enhanced perception and insight during the experience even though, by ordinary medical criteria, they have no brain function. They commonly report leaving their bodies and viewing events from above the scene. Interestingly, they report being able to see and hear in some way, but they generally don't report smell, taste, or bodily feelings.

A life review is more common among survivors of clinical death after accidents than among NDErs who have suffered prolonged illness. We don't really have enough information yet from individual accounts to suggest reasons for that.

There are many variations on the theme of the classic near-death experience as well. Some experiencers see a silver cord, and most have a powerful sense of bliss (although occasionally they can have a sense of terror). They generally sense that the light they see is a transition from death to new life and a sense of timelessness.

What Do Researchers Say?

Overall, the experiences are positive, with intense feelings of love, warmth, and acceptance. NDErs who pass through the tunnel don't usually wish to return to their bodies. But they are encouraged to do so, often because they are told that they still have things to do in this life. Tulane University psychologist Marilyn A. Mendoza, a specialist in grief counseling, succinctly expresses what many counselors have noted: "Perhaps the most common after-effect of an NDE is the loss of the fear of death and a strengthened belief in the afterlife. There is typically a new awareness of meaning and purpose in experiencers' lives. A new sense of self with increased self-esteem is reported."[16] That effect shows up in research studies too.[17]

Leeds Beckett University psychologist Steve Taylor, author of *Spiritual Science* (2018), offers a striking fact about the depth of the transformation:

> It's remarkable that one single experience can have such a profound, long-lasting, transformational effect. This is illustrated by research showing that people who have near-death experiences following suicide attempts very rarely attempt suicide again. This is in stark contrast to the normal pattern—in fact, a previous suicide attempt is usually the strongest predictor of actual suicide.[18]

That is indeed a significant finding. Some might argue that people who recall NDEs are overstating their newfound commitment to a different way of seeing life. But when suicidal people stop attempting suicide, they have clearly undergone a concrete and highly significant behavior change. Generally, the best predictor of any future behavior is past behavior. So whatever happened to them is apparently not just imagination; it changes future behavior in a significant way.

Some Near-Death Experiences Are Frightening

While the great majority of NDEs are pleasant, experiencers have reported unpleasant or frightening ones.[19] Researchers debate how common the negative ones are. It's hard to say because experiencers may be reluctant to discuss them. However, distressing NDEs can also result in insight and personal growth. One young woman whose NDE left her crushed by a sense of inadequacy nonetheless heard a voice telling her not to be too hard on herself, at which point she sensed she was returning to her body:

> I ended up accepting myself for the person that I was... knowing that I was loved and loveable... I've never really succeeded because my, actually my friends were pretty screwed up people... it's just accepting them and accepting the best possible solutions as far as day-to-day relationships went.[20]

Researchers Nancy Evans Bush and Bruce Greyson have also found that such experiences can have an upside, quoting one account: "I was being shown that I had to shape up or ship out, one or the other. In other words, 'get your act together,' and I did just that."[21]

The aftermath of NDEs is not all roses anyway, though, as Greyson explains:

> Some experiencers have difficulty reconciling their NDEs with their religious beliefs. Some find it hard to resume their old roles and lifestyles, which no longer have the same meaning, or to communicate to others the impact of the NDE. Some experiencers report anger at still being alive—or at being alive again.[22]

The Human Mind Beyond Death

Perhaps as a result, NDEs do not necessarily contribute to peace and harmony in relationships. The marriages of NDE experiencers are more likely to end in divorce, due, in the view of researchers, to incompatible value systems.[23]

Near-Death Experiences Span Both History and the Globe

Near-death experiences have been reported for thousands of years. Gregory Shushan, author of *Near-Death Experience in Indigenous Religions* (2018), notes their global reach:

> NDEs have been popularly recognised in the West since the mid-1970s, but people from the largest empires to the smallest hunter-gatherer societies have been having them throughout history. Accounts are found in ancient sacred texts, historical documents, the journals of explorers and missionaries, and the ethnographic reports of anthropologists. Among the hundreds I've collected are those of a 7th-century BCE Chinese provincial ruler, a 4th-century BCE Greek soldier, a 12th-century Belgian saint, a 15th-century Mexica princess, an 18th-century British admiral, a 19th-century Ghanaian victim of human sacrifice, and a Soviet man who'd apparently killed himself but was revived during resuscitation experiments. NDEs can happen to followers of any religion, and to those of none.[24]

In his research, he found over seventy Native American NDE accounts from the sixteenth through the nineteenth centuries and thirty-six from the Pacific Islands as well.[25] They have usually been treated as mystical experiences or otherworldly journeys.[26]

Atheists can have near-death experiences too. Howard Storm, an atheist art professor at Northern Kentucky University, had a hellish NDE amid a near-fatal attack of peritonitis.[27] He subsequently left his university post and attended seminary.[28] British philosopher A. J. Ayer (1910–1989) had an NDE in 1988. Although he did not revise his atheist views, he is reported to have become a much nicer person to deal with,[29] which is consistent with findings that NDE experiencers often undergo significant changes in behavior.

It's difficult to prove an NDE to those who have not had one, but profound, consistent, long-term personal change is not easy. When it occurs, we should infer that a person has experienced something significant, even if we don't clearly understand what happened.

But there is another line of evidence too.

Seeing Things That Others Can Verify

People undergoing a near-death experience are sometimes aware of events that they should not be able to observe under the circumstances, as we saw with Pam Reynolds. This type of NDE is called *veridical*, meaning that the information reported is later confirmed.

Bruce Greyson, author of *After* (2021), was an agnostic psychiatrist decades ago, skeptical that the mind could be detached from the brain. But then a young woman, rescued from suicide, told him that she had seen a spaghetti stain on his tie during an out-of-body experience. There was indeed such a stain—and he had gone to some trouble to conceal it from colleagues.[30]

Unable to just forget that embarrassing but undeniable episode, Greyson began to study near-death experiences from a science perspective. He went on to develop an NDE scale for assessment that met the standards for professional diagnostics.[31] Incidentally, he remained a religious agnostic. To him, NDEs are a matter of following the evidence.

In his book, he also discusses a revealing veridical experience that happened to another doctor, a surgeon. A truck driver undergoing quadruple bypass heart surgery—fully anesthetized, with eyes taped shut—claimed that he had "come to" during the procedure. He experienced looking down at his own body as the doctors prepared to operate. Specifically, he recalled the surgeon waving his elbows in the air "as if he were trying to fly."[32] When he brought this up later, the embarrassed surgeon did not want to discuss it. So Greyson, who was curious about the matter, met with the surgeon to find out more about the patient's claim.

The surgeon admitted to Greyson that he did in fact wave his elbows in the air. Having been trained in Japan, he followed a cultural practice of pointing with his elbows so as to avoid any risk of contamination of his gloved hands.

One of my own colleagues similarly encountered a young child who underwent complex skull surgery. The boy described his own operation in meticulous visual detail, even though he was under general anesthesia and had his eyes and face covered during the procedure. Significantly, that neurosurgeon had a specific way of doing the procedure; thus, technical details the child reported reflected his individual preference. The information was not publicly available and neither the family nor the child had been told about it in the detail that the child reported. The family was so shocked by the child's account that they asked why my colleague hadn't used anesthesia for the surgery![33] The child's mind appears to have been functioning quite independently of his brain.

NDE researchers define these veridical perceptions as "any perception—visual, auditory, kinesthetic, olfactory and so on—that a person reports having experienced during one's NDE and that is later corroborated as having corresponded to material consensus reality."[34] Near-death experiences have been subjected to extensive scientific analysis on this basis over the past half century. In 2009, Janice Holden and colleagues reported 107 cases in the scientific literature of NDE patients who had veridical experiences involving observations and events that could be corroborated in the way we are describing.[35]

Seeing Things That Others Do Not Know

In *After*, Greyson identifies another type of information that can result from NDEs: seeing a person who has died but is not known to have died. He offers an example from one of his papers:

> A young nine-year-old boy named Eddie was seriously ill in a hospital. Recovering from a thirty-six-hour fever, Eddie immediately told those in the hospital room that he had been to heaven, recounting seeing his grandfather, an aunt, and an uncle there. But then his startled and agitated father heard Eddie report that his nineteen-year-old sister Teresa, away at college, was in heaven too, and she told Eddie that he had to return. But the father had just spoken to Teresa two days prior. Checking with the college, the father found out that his daughter had been killed in a car accident the previous day.[36]

The college had not been able to reach Eddie's parents immediately, probably because they were away with Eddie.

Greyson notes, "None of the NDEs in our collection involved an experiencer mistakenly thinking a person still alive had died. NDEs in which the experiencer meets—and is surprised to see—a loved one they hadn't known had died are not common, but they do occur."[37]

Increasing Interest in Recent Years

Naturally, as modern medicine brings more people back to talk about their experiences, public interest in verified NDEs has grown with the body of evidence. The 2023 Angel Studios documentary *After Death* was a box-office hit,[38] despite being panned by well-known critics[39] as "lacklustre," "fairly limp," and a "repetitive slog." What the critics did *not* say is perhaps much more important: They did not point to science evidence that easily refutes the film's claims.

However, because near-death experiences challenge the view that the mind (the human soul) is simply the physical functions of the brain, there are many efforts to debunk them. Addressing some of these efforts will give us a clearer understanding of the relationship between the human soul and the body.

CHAPTER 6

The Skeptics' Turn at the Mic

YEARS AGO, I WAS a bit skeptical about near-death experiences myself. Other science evidence for the soul that I studied was clearly compelling—Wilder Penfield's brain mapping and Roger Sperry's studies of split-brain patients are solid science of the highest quality. Penfield was the greatest neuroscientist in the neurosurgical profession, and Sperry won the Nobel Prize for his research. The science of near-death experiences, on the other hand, seemed less solid.

I've never had a patient report a near-death experience. But that in itself doesn't prove anything. Neurosurgeons don't hear about many near-death experiences because when our patients are sick unto death, their brains are generally so damaged that they cannot afterward remember or communicate their experiences. Cardiologists and anesthesiologists see many more patients with near-death experiences, because their patients, upon returning to life, usually have good brains and can readily remember and recount what happened to them when they died.

The Skeptics' Turn at the Mic

The anecdotal reports by thousands of people who had died, been resuscitated, and then reported remarkable otherworldly experiences were astonishing. But they were quite different from traditional scientific experiments or studies. You can't predict near-death experiences, and obviously you can't deliberately evoke them to study under controlled conditions. However, some reports, such as that of Pam Reynolds in chapter 5, are very well authenticated. The science of these veridical experiences—perhaps about 20 percent of near-death experiences—was reasonably good. But still, the overall science of near-death experiences could be described as a vast collection of anecdotes, some of which can be partially verified, which is hardly ideal science.

Inferences to the Best Explanation

But science is always about making provisional explanations—inferences to best explanation. The truth of a theory of near-death experiences, such as that they represent genuine spiritual experiences, always depends on its credibility compared with that of alternative theories. My opinion about the science of NDEs changed dramatically as I studied experiences recorded in the medical literature. Thousands are now reported—and when I compared the evidence for their reality to the evidence that they were merely the consequence of brain chemistry or wishful thinking and so on, the theory that they are genuine spiritual experiences looks strong.

For one thing, while I have not personally had a patient who reported a near-death experience, after over seven thousand brain operations, I do have a lot of experience with the conditions that skeptics claim are their true cause.

To take just one example (we discuss many in the remainder of this chapter), many materialist scientists attribute near-death experiences to low brain oxygen levels that cause brain dysfunction and unusual experiences. But I've treated thousands of patients with low brain oxygen levels. Low oxygen causes extreme confusion, fear, disorientation, diminishment of perception (vision fades to black, then hearing fades), and severely impaired memory. This is completely unlike the startling mental clarity and beautiful scenes that near-death experiencers report. Lack of brain oxygen destroys the mind. Near-death experiences enhance the mind in remarkable ways—many patients say that it seems like ordinary life is living in a darkened room, and we don't see with clarity until we die.

I have come to believe that the theory that near-death experiences are real is very good science, because although the experiences are anecdotal by nature, multiple lines of evidence point to the conclusion that they are real spiritual experiences. The real issue is that NDEs just don't fit into the ideological framework of current science—that we are just meat and we lack souls.

By contrast, my experience of forty years of clinical practice and tens of thousands of patients in very extreme circumstances teaches me that none of the materialist explanations we hear is the least bit credible. To see what I mean, let's take a closer look at those skeptical theories.

Celebrity astronomer Carl Sagan (1934–1996), facing death, spoke for many when he said, "I would love to believe that when I die I will live again, that some thinking, feeling, remembering part of me will continue. But as much as I want to believe that, and despite the ancient and worldwide cultural traditions that assert an afterlife, I know of nothing to suggest that it is more than wishful thinking."[1] He pioneered one of the efforts to explain away near-death experiences.

The Skeptics' Turn at the Mic

NDEs as Memories of Our Own Births

Sagan popularized the idea that traveling down a tunnel into the light is just a "birth memory":

> Every human being, without exception, has already shared an experience like that of those travelers who return from the land of death: the sensation of flight; the emergence from darkness into light; and experience, in which, at least sometimes, a heroic figure can be dimly perceived, bathe the radiance and glory. There is only one common experience that matches this description. It is called birth.[2]

As a number of critics have pointed out, his hypothesis ignores the obvious difference between the typical NDE experience of peace and joy and the shock felt by a screaming newborn in the delivery room.

Sagan's theory doesn't make sense for many reasons. When a baby exits the womb, the top of the head usually emerges first. Thus the baby is looking backward and would not see anything resembling a tunnel. The light described by near-death experiencers is consistently described as warm and loving. The light of the delivery room, if viewed by a newborn at all, is just the opposite—painfully bright. Everything is suddenly much colder than the mother's womb. Anyone who has observed a typical birth knows that it is a time of great distress for the baby, not at all a peaceful and reassuring moment.

And what about babies who are born by the less stressful method of cesarean section? Psychologist Susan Blackmore (who does not believe in the survival of the mind/soul after death) conducted a study. She compared the tunnel and out-of-body experiences of near-death experiencers who had been born by vaginal delivery with those who had been born by cesarean section. No difference was found between the two groups.[3] So birth memories could not have been the cause of the NDEs.

Sagan's birth memories hypothesis is implausible for other reasons as well. Newborns have very limited abilities to perceive what's going on or to form memories.[4] Think about it: What do you remember of your own birth or even of the first few years of your life? A bigger problem for his thesis is that NDErs who have the tunnel experience may, like Pam Reynolds, have no brain activity at all. In that case, conventional memory is not even possible. Sagan's suggestion mainly shows how improbable a materialist explanation can be—and still be preferred to a nonmaterialist one.

Is It Just Wishful Thinking?

The RationalWiki encyclopedia project informs us that "wishful thinking may generate experiences confirming what a subject wants to believe and brain damage may prevent a subject recognizing that the experience was a dream."[5] The obvious problem with such a claim is that clinical brain death, where confirmed, precludes any mental activity at all, including wish fulfillment. But suppose it were possible. As prominent Allen Institute neuroscientist Christof Koch points out,[6] NDEs are no more likely to occur to people who believe in an afterlife than to those who don't. Western and non-Western experiencers tend to have the same core experiences.[7] Those experiences don't necessarily confirm personal beliefs either. For at least one Israeli man who became the subject of a research paper, the near-death experience was very upsetting on that account.[8]

There are three other reasons to doubt RationalWiki's explanation. As radiation oncologist and near-death researcher Jeffrey Long notes, some NDE experiencers provide accurate accounts of events that they themselves did not expect to occur. That, he says, "argues against

NDEs as being a result of illusory memories originating from what the NDErs might have expected during a close brush with death."[9] There is also considerable evidence of NDEs as profound and life-changing,[10] which, again, is not what we would expect of mere wishful thinking.

Sam Parnia, author of *Lucid Dying* (2024), also notes that when NDE experiencers feel that their lives are under review, "They don't review their lives based on what people strive for, like a career, promotions, or an amazing vacation. Their perspective is focused on their humanity. They notice incidents where they lacked dignity, acted inappropriately towards others, or conversely, acted with humanity and kindness."[11] For example, Nottingham Trent University psychologist David Wilde reports that a woman he interviewed told him that "when her heart stopped, she felt that she was in a dark void where she reflected on everything bad she had ever done in her life before hearing a voice telling her not to be too hard on herself." As a result, she sensed the need for a new beginning and changed course in life, becoming an interfaith minister and counselor.[12] That probably wasn't wishful thinking; there is no evidence that she had wished for any of it.

Does the Trauma of Impending Death Cause People to Imagine Near-Death Experiences?

Psychiatrists have wondered whether a sense of unreality or detachment from one's body (*depersonalization*)[13] could produce NDEs. The problem with that explanation is that the NDE experience is profoundly different from depersonalization. Psychiatrists Glenn Gabbard and Stuart Twemlow have identified important differences between NDEs and depersonalization.

Depersonalization is generally experienced *in the body* and is associated with intense anxiety and dysphoria. That's quite unlike the generally very pleasant (or at least constructive) out-of-body NDE experiences.[14] It's also significant that psychologist Harvey Irwin found that NDE experiencers are not, as individuals, predisposed to depersonalization experiences.[15] In any event, they feel more in touch with their real selves during NDEs than before.

For example, cardiologist Michael Sabom reports the account of one NDE experiencer: "I've had a lot of dreams and it wasn't like any dream that I had had. It was real. It was so real. And that piece, the piece made the difference from a dream, and I dream a lot."[16] Another NDEr confessed, "That was real... I've lived with this thing for three years now and I haven't told anyone because I don't want them putting the straitjacket on me... It's real as hell."[17] Sabom, incidentally, admits in the Angel Studios film *After Death* that when he first heard a patient recounting an NDE, he thought it was "hogwash." But after the third such account, he began to listen more carefully.

Could It Be Just Imagination?

Some skeptics suggest that near-death experiences are merely imagination[18]—they could be educated guesses based on watching medical television dramas. Is that likely? Sabom found that NDE experiencers who had suffered cardiac arrest generally showed more knowledge of what had happened in the emergency room in which they had been treated when they were clinically dead than patients who had a cardiac arrest but who didn't have a near-death experience.[19] A similar study by Penny Sartori found the same thing.[20] A 2014 study noted that awareness of what was happening during cardiac arrest, while documented, is not something that current science can account for: "We reached the inevitable conclusion that patients experienced all the aforementioned NDE elements during the period of their cardiac arrest, during the total cessation of blood supply to the brain. Nevertheless, the question how this could be possible remained unanswered."[21]

Were the NDE Experiencers Really Semiconscious?

Some critics have suggested that NDEs are a consequence of ordinary perception while the dying patient is still semiconscious. One problem with that explanation is that patients who are semiconscious due to anesthesia report *hearing* medical activity going on around them but not *seeing* it.[22] NDE experiencers both see and hear. Sabom recounts a patient who had experienced both perception during a medical emergency *and* perception during an NDE. The patient reported the difference:

> I didn't see nothing [during the non-NDE medical emergency] I just heard. This other time [the NDE] with the cardiac arrest, I was looking down from the ceiling and there were no if's and's or buts about it.[23]

Semiconscious perception also does not account for recorded instances of knowledge of conditions in the operating room that would just be unavailable to ordinary perception, such as Pam Reynolds experienced. It seems quite clear that NDEs enable a wholly different type of perception from the occasional awareness that occurs under anesthesia.

When All Else Fails... Try Darwinian Evolution

Costanza Peinkhofer and colleagues have suggested that NDEs are an evolved human example of a quality that some animals possess, *thanatosis*. Opossums are famous for this. They can "play dead" to discourage predators that prefer live prey.[24] And they really do look dead. The trouble is, the proposed explanation doesn't account for much. Animals whose metabolisms automatically shut down as a defense mechanism can't help us understand much about the explicitly spiritual nature of human NDEs.

Could NDEs Be Caused by Hallucinations Generated by the Death Process?

Many proposed explanations for near-death experiences are based on physical events during the dying process. For example, the process might lead to hallucinations. Hallucinations are generally fragmented, disturbing, and subrational experiences. We should not expect those generated by the death process to be any different. And yet near-death experiences are *very* different.[25] Here's a typical account from the NDE literature:

> I went into a dark place with nothing around me, but I wasn't scared. It was really peaceful there. I then began to see my whole life unfolding before me like a film projected on a screen, from babyhood to adult life. It was so real! I was looking at myself, but better than a 3-D movie as I was also capable of sensing the feelings of the persons I had interacted with through the years. I could feel the good and bad emotions I made them go through.[26]

Nothing about this account suggests derangement. Claims about hallucination also cannot account for the way near-death experiencers know what is going on around them when they are clinically dead. Nor can they explain another feature to which researcher Jeffrey Long draws attention: "Life reviews may include long forgotten details of their earlier life that the NDErs later confirm really happened. If NDEs were unreal experiences, it would be expected that there would be significant error in life reviews and possibly hallucinatory features."[27]

Also, in both child and adult near-death experiences, the relatives they encounter are almost always *dead* ones, not living ones.[28] It is not clear why mere hallucinations would selectively image deceased persons. As Long notes, dreams and hallucinations are much more likely to conjure up living persons in recent memory.[29]

Could Near-Death Experiences Really Be Temporal Lobe Seizures?

Some have claimed that when pioneering neurosurgeon Wilder Penfield stimulated the temporal lobes of patients undergoing brain surgery while awake, he prompted NDE-like experiences.[30] However, that is just not likely. First of all, these seizures have a very well-characterized pattern that neurologists deal with on a regular basis. They can include fear, joy, déjà vu, strange odors or tastes, a roller-coaster sensation or failure to recognize familiar people. But that is hardly what NDE experiencers report. Furthermore, it can't be emphasized enough that near-death experiences can happen when there is no circulation of blood to the brain. There is almost certainly complete electrical silence rather than the disordered electrical activity of a seizure.[31] In short, there is nothing that resembles a near-death experience as described in the literature.

What About Shortage of Oxygen to the Brain?

Could brain oxygen shortage (hypoxia) produce near-death experiences, as some claim?[32] Generally, hypoxia produces intense agitation, fear, and confusion, as well as lethargy, irritability, and inability to concentrate, not the peace and clarity of NDEs. The Near-Death Experience Research Foundation (NDERF), which has collected evidence of over five thousand such experiences worldwide since 1998 (as judged by Greyson's NDE scale),[33] asked in a survey, "How did your highest level of consciousness and alertness during the experience compare to your normal, everyday consciousness and alertness?" Over 74 percent of 1,122 NDErs surveyed said they had "more consciousness and alertness than normal"; only 5 percent said they had less.[34] So what they experienced is the opposite of what we see in typical low-oxygen states of the brain.

Anyway, the oxygen levels of dying patients who have near-death experiences have turned out to be higher than those of patients who don't have them. Blood carbon dioxide levels were the same.[35] So NDEs are not really about shortage of oxygen to the brain.

Could NDEs Be Caused by High Carbon Dioxide (CO_2) Levels in the Bloodstream?

The claim that high levels of CO_2 (hypercarbia) explain many near-death experiences[36] faces a big problem: Blood carbon dioxide levels can have no effect on brain function if the brain is clinically dead. And when tested, carbon dioxide levels in the blood of patients who had near-death experiences during resuscitation were not high.[37] When hypercarbia occurs in more ordinary situations, it features bizarre hallucinations, brightly colored patterns, and perceptual confusion, which, again, is quite different from the accounts of most NDE experiencers.

Could NDEs Be Created by Chemicals Produced Naturally in the Brain?

Many skeptics claim that near-death experiences are due to neurochemicals such as endorphins that are generated by the dying brain or to drugs administered during a resuscitation attempt.[38] But, once again, the problem is that it is easy to distinguish between the effects of NDEs and those of chemicals.[39] Drugs and brain chemicals generate varied, inconsistent, and idiosyncratic mental states, with bizarre visual perceptions such as cobwebs, honeycombs, and lattices, not the calm visions of typical NDEs.

Some skeptics have suggested that the specific peptides endorphin and enkephalin may be released in the brain during the dying process and that their stress-reducing effect may account for near-death experiences.[40] But the timing doesn't work. The course of near-death experiences is often very quick, but endorphins and enkephalins have relatively long half-lives; they decay slowly over minutes or hours. Therefore, the experiences they induce would tend to dissipate only very slowly. In near-death experiences, pain and uncomfortable feelings return instantly when the soul reenters the body.

Bruce Greyson looked into the suggestion that drugs administered to the patient were the cause of NDEs. He found the opposite of what skeptics might expect: "The more drugs people are given as they approach death, the less likely they are to report a near-death experience. So drugs and lack of oxygen are not causing NDEs. They may in fact, repress having an NDE."[41]

Efforts to tie NDEs to chemical imbalances can be highly inventive. In 2019, *Scientific American* reported an attempt to equate near-death experiences with recreational drug highs, *based only on language use*.[42] We may as well infer that lung cancer and tuberculosis have a common cause. After all, sufferers from both diseases complain to their doctors about coughing, shortness of breath, chest pain, and unexplained weight loss. But that is not a cause-and-effect relationship. It is all effect, no cause. Medicine should focus on causes, not merely effects.

Perhaps the most sophisticated chemical theory proposed to explain away near-death experiences is the ketamine theory.[43] Ketamine is a drug that blocks amino acid NMDA receptors in the brain. It suppresses sensory input and activates the central nervous system, producing an effect similar to sensory deprivation. Some ketamine reactions do indeed resemble near-death experiences—in a limited way.

Ketamine is more likely to cause what doctors call an "emergence" reaction. That just means a state of great confusion, which the patient experiences as very unpleasant. We do not currently know if the brain even produces ketamine. Any such acute dysfunction of the brain during the dying process would result in confusion, not in the usually reported clarity.

Are NDEs Just What Happens to a Dying Brain?

Researcher Susan Blackmore offers a simple explanation. Maybe the tunnel seen in near-death experiences is the product of retinal disturbances. That is, retinal cells in the eye, caught up in the process of dying, might image a tunnel. She proposes that peripheral vision blacks out before central vision does.[44] Perhaps, but the tunnels described in near-death experiences are clearly visible and highly detailed, which would not be expected in tunnel vision caused by the death of retinal cells during the dying process. Also, many near-death experiences *don't* involve a tunnel. Tunnels are seen more commonly by Western near-death experiencers than by Asian experiencers, says Chris Carter, author of *Science and the Near-Death Experience*.[45] The "tunnel" may just be a way of describing an otherwise hard-to-describe experience.

More ambitiously, Blackmore offers a broader *dying brain hypothesis*.[46] In that interpretation, near-death experiences result from a variety of physiological processes underway in the brain during the death process. She asserts that "the brain builds an image of reality to compensate for disintegration." The difficulty with this theory is the obvious one. A brain disintegrating in the process of dying is hardly able to build any image of reality at all, let alone an astonishingly clear, compelling, and often beautiful one.

The bottom line is that the dying brain is disorganized and failing, but near-death experiences are generally highly organized and peaceful. That rules out all the physiological explanations currently on offer.

The Weaknesses of Materialist Explanations in General

As Steve Taylor has pointed out in *Psychology Today*, "NDEs have never been satisfactorily explained in neurobiological terms…All of these theories have been found to be problematic."[47] One weakness of such explanations is that they try to account specifically for selected aspects of the near-death experience. But they fail to account for the complete experience. The ways they account for the narrow aspect of NDEs that they focus on invariably conflict with all the other evidence. That is why no skeptical explanation provides a credible account of NDEs.

The AWARE Trials

In 2014, NDE researcher Sam Parnia reported the results of a *prospective* study, the AWARE study (AWAreness during REsuscitation), that began in 2008, involving fifteen hospitals in the United States, the United Kingdom, and Austria.

A prospective study is planned prior to the cases studied, which are then collected according to the rules of the study. By contrast, a *retrospective* study looks at cases that occurred before the study was planned. That means that the original data was collected without a research format.

Prospective studies are generally considered better science, because a well-thought-out study design can filter out factors that might complicate the findings. But practically speaking, NDEs are challenging to study prospectively. Life-threatening medical emergencies are not a promising environment for research studies. They cannot be scheduled in advance, they are not predictable, and when they do occur, you can either save the patient or you can't. Sometimes the effort includes hectic and desperate efforts to resuscitate a person who is clinically dead.

With these constraints in mind, the researchers reported on 2,060 patients who had suffered cardiac arrest, among whom 140 survivors could be interviewed. Only 101 were able to complete detailed interviews. Nine of them had near-death experiences. Two showed awareness of their environment at the time of the experience, and one of these accounts was certified as accurate by personnel who were in the room at the time of the resuscitation.[48]

In particular, one patient reported hearing an *automated* voice saying, "Shock the patient, shock the patient."[49] Medical records showed that an AED (automated external defibrillator) had been used. The AED barked just the sort of automated instructions that the patient heard and was not likely to otherwise know about. His heart was defibrillated because it had stopped and his brain was without blood flow. Yet he heard the automated voice from the device.

This patient's experiences were validated and could be timed because the automated voice of the defibrillator occurred at a precise time that was recorded by doctors and nurses during the cardiac arrest. Parnia concluded:

> This is significant, since it has often been assumed that [these] experiences... are likely hallucinations or illusions, occurring either before the heart stops or after the heart has been successfully restarted, but not an experience corresponding with "real" events when the heart isn't beating. In this case, consciousness and awareness appeared to occur during a three-minute period when

there was no heartbeat. This is paradoxical, since the brain typically ceases functioning within 20–30 seconds of the heart stopping and doesn't resume again until the heart has been restarted. Furthermore, the detailed recollections of visual awareness in this case were consistent with verified events.[50]

The AWARE II trial at twenty-five hospitals in the United States and United Kingdom features hospitalized patients as well as 126 other survivors of cardiac arrest who recalled NDEs, to provide comparative information. It has so far found that perhaps one in five people who survive cardiopulmonary resuscitation (CPR) after cardiac arrest may describe lucid experiences.[51] It's not clear, of course, how many of them were, like Pam Reynolds, essentially dead at the time. But something seems to be happening for which the view that the mind dies with the body does not account.

The AWARE trial authors sum up their position, based on years of research:

> The study authors conclude that although studies to date have not been able to absolutely prove the reality or meaning of patients' experiences and claims of awareness in relation to death, it has been impossible to disclaim them either. They say recalled experience surrounding death now merits further genuine empirical investigation without prejudice.[52]

Considering the number of objections that must be overcome, we've come a long way when we can just discuss the topic honestly from a medical science perspective.

Evidence for the Reality of Near-Death Experiences

Many studies point very strongly to the reality of near-death experiences. Efforts to explain them away do not really work. They have been documented in cultures throughout the world for thousands of years. From a research perspective, we could say that the retrospective evidence for NDEs is *massive*. Some near-death experience patients, like Pam Reynolds, have very well documented complete absence of brain function.

It's also significant that religious beliefs and previous knowledge do not necessarily predispose a person to a near-death experience. They are stable across cultures and often life-changing. All of these characteristics are hallmarks of an objective experience. As Jeffrey Long told an interviewer in 2023, "I've read brain research and considered every possible explanation for NDEs. The bottom line is that none of them hold water. There isn't even a remotely plausible physical explanation for this phenomenon."[53]

What Do Near-Death Experiences Tell Us About the Human Soul?

First, verified NDEs confirm that there is an immaterial aspect to the human person—call it mind or soul—that *survives the death of the brain*. The sheer number and variety of materialist explanations that attempt to get around that are a symptom of the weakness of a materialist position on this topic. It's got to the point where even skeptics like Susan Blackmore and Michael Shermer disparage the usual routine dismissals of NDEs as crude and unhelpful in the face of a genuine mystery.[54] As Bruce Greyson has said, "Far from leading us away from science and into superstition, NDE research actually shows that by applying the methods of science to the nonphysical aspects of our world, we can describe reality much more accurately than if we limit our science to nothing but physical matter and energy."[55]

All the materialist attempts to explain NDEs away that we've encountered involve implausible scenarios. Were it not for a strong materialist ideological bias among some scientists, none of them would be taken seriously in the scientific community. However, as philosopher Gary Habermas points out, long-standing biases can outweigh evidence: "It often appears that the major underlying conflict in these matters is not primarily about evidence but more about a momentous clash between worldviews... If this is accurate, then it seems that even strong evidential considerations are less likely to change minds."[56] We must then focus on ensuring that the information is available to those who are willing to consider it.

But Here Is the Next Really Big Question...

Near-death experiences catch the mind, the human soul, in the act of surviving the death of the brain. Does that survival prove that the soul is immortal? What if the mind survives only temporarily after the brain ceases to function? That may seem unlikely. But to rule it out, we must demonstrate that the soul belongs to a class of things that are, *by their very nature*, immortal. There are, we will show, some pretty good reasons to think that the soul, unlike the body, not only does not die, but *cannot* die.

CHAPTER 7

Immortality of the Soul Is a Reasonable Belief

I DISLIKE FUNERALS, ESPECIALLY funerals of my patients. Surgeons don't like failure. A few years ago, I walked into the room in the funeral parlor where Jenny was in repose. She was only ten years old, and her short life was beset with illness. She was born with severe mental handicaps and a persistent accumulation of spinal fluid—essentially water—in her brain. Scores of brain operations had kept the water from causing further brain damage, but ultimately she was so weakened by her illnesses that she died from pneumonia.

Her mom, dad, sister, and brother were seated in the front row. I expressed my condolences. They were grateful I came. I went to the back of the room to find a seat, but there weren't any, so I stood in the back against the wall.

The room was packed with at least a hundred people, and scores more were spilling out into the hallway. It was the best-attended funeral I had ever seen. Most of the mourners were people from the hospital—nurses, clerks, aides, x-ray technicians—as well as neighbors

Immortality of the Soul Is a Reasonable Belief

and friends of Jenny's family. We had gotten to know her parents quite well, due to Jenny's many operations, hospitalizations, and trials.

The pastor was an old friend who helped me in my journey from atheism to Christianity. He had a way of explaining difficult things, but I didn't envy him his job that day. What can you say to a family and a room packed with fellow mourners about the suffering of a little girl who never sat up, never learned to talk or feed herself, and went through scores of major operations? I thought he would probably say that at least in death her suffering, like her life, was extinguished.

He said something quite different:

> Look around you. There are more people here at her funeral than will attend the funeral of anyone in this room. Esteemed professors, experienced nurses, hardworking aides and neighbors are unlikely to have this many people gather to mourn their passing.
>
> In this sense, her life brought much good to the world. I don't mean to belittle her suffering and the suffering of her family—their suffering was intense—but suffering can be redemptive, and the redemption her suffering gave to us lives on in this room...
>
> You might think that her accomplishments were different from those of healthy people, that she was a mere vehicle for what she accomplished, but, unlike healthy folks like us, she couldn't really take credit for it. But that would be to misunderstand our own lives. We are given our opportunities and our talents, given everything we have and do, given every hair on our head, given life itself...
>
> Jenny's life and accomplishments are different from ours but are no more or less dependent on God than ours

are. Our very existence is in His hands from moment to moment, and every bit of good we do depends on Him. What's more, her remarkable accomplishments were done with the humility that only a severely disabled child really knows. Pride is an affliction of health. Disabled children are gracefully free of it.

Now Jenny is whole, fully healed, running and laughing in the fields of an immeasurably more beautiful land, loved there as she is here, but loved not just by us fallible mortals, but by Love Himself, Who she sees now face-to-Face. We celebrate her birthday and her restoration to health in paradise, and we thank her for her gift to us—her gift of love that binds us all together today.

There were muffled sobs as we all took in this new way of looking at the life of this little girl.

The pastor's homily really made me think. Is it possible that Jenny, who lived with crippling disabilities and whose dead body lay in repose in front of me, is yet alive, completely healthy, running and frolicking in an empyrean valley? How could eternal life even be possible—doesn't bodily death end our existence once and for all? At this point in my intellectual journey, I had become sufficiently suspicious of blithe materialist cant like "All life ends at physical death" that I approached the question of the immortality of the soul with an open mind.

Could the pastor be right? Let's take a look at the logic and the evidence for immortality of the soul.

A Perennial Belief

Over a hundred thousand years ago, in Israel, some forty individuals were buried in pits. The pits also held a shell or red ochre. One pit held part of a deer's skull and antler.[1] Archaeologist Avraham Ronen (1934–2018) has commented, "The gifts offered to the dead might be alluding to some kind of religious belief in rebirth and afterlife."[2]

Similarly, a small child buried seventy-eight thousand years ago in Panga ya Saidi, a cave site on the Kenyan coast, was found in a flexed, fetal position.[3] Did the mourners think the child might be born again somewhere?

We'll never know, but it would surely be no surprise. The idea that the human soul survives death is perennial—people seem to have always had it—and it is almost universal. It is actually not so much a hope as an expected fate. Survival, after all, could mean paradise, nirvana, limbo, reincarnation—or torment. Any fate at all is in sharp contrast to a cornerstone belief of current neuroscience that the mind is simply what the brain does and that it therefore disintegrates at death. Thus an immortal human soul cannot really exist.[4] From what we have seen so far, there is every reason to believe that the soul does exist and that we can know some things about its nature. One of those things we can know is the reason why the soul is immortal.

What Do We Mean When We Say That a Human Being Dies?

We can defend the immortality of the human soul by pointing to verifiable accounts of near-death experiences. But in themselves, such experiences only tell us that the mind can continue to function when the brain is clinically dead. That could mean one of two things: Perhaps it means that the mind can survive without the brain but only for a period of time. Even physical death, as we are learning, does not occur all at once; it takes literally days for everything in a human body to die.[5]

But what if the mind can function independently of the brain forever because it belongs to a class of entities that are immortal by their very nature? What if the soul not only does not die but *cannot* die? We've talked about the way that pure concepts like justice or the number 7 can't die. That's because they do not begin or end. Yet in the world around us, every living thing—plant, animal, human, or whatever—is physical, which means that it begins and ends physically. We need to be clear about how the human soul is different.

When we say that living things have *souls*, we are referring to all the tightly integrated processes that work together to make them what they are. Let's look again at what exactly happens when a sparrow dies as opposed to what happens when a human being dies. Death is one form of *disintegration*—the dis-integration of the sparrow's form (its soul) from its matter. The form vanishes. The matter, left to itself, is eagerly recycled by countless still-living forms. When they are done with it, nothing like a sparrow is left.

When a human being dies, everything that is mortal about that person also disintegrates in the same way. But everything that participates in immortality—ethical choices for good or evil, for example—remains. Those choices cannot, by their nature, die, any more than the principles of justice or mercy on which they depend can die. That's because they are not physical and do not disintegrate or end.

Some may think that immortality of the rational soul is not possible in the physical world in which we live. But they are mistaken.

Even a Candle Does Not Just "Go Out"

People who think we are completely annihilated at death have a favorite image: "Where is the flame when the candle goes out?" The annihilationists are being careless here. They assume that the physical flame just disappears. Actually, it doesn't. As Boston College philosophy professor Peter Kreeft observes, "The light does not go out; it goes up. It is still traveling through space, observable from other planets."[6]

Immortality of the Soul Is a Reasonable Belief

Yes, that's true. In the material world, energy is neither created nor destroyed. But it is often transformed from one state to another. In fact, nothing in this universe simply dissipates; it is always transformed. The immaterial world is similar. An immaterial reality like the human soul may be transformed into a different immaterial reality, but it cannot simply be annihilated.

Some people will argue that immortality does not make sense because we cannot locate the soul after the body has died. Well, let's think about that. Could we locate the soul before death? We know about the soul—both the mortal aspects like movement and emotion and the immortal aspects like reason and free will—only by observing the behavior of the living body. Apart from physical behavior, the soul does not actually have a location.

Kreeft, author of *Socrates' Children* (2019), suggests that the human body, far from creating the soul, may even act as a restraining force on it: "Thus when the brain dies, more rather than less consciousness occurs: the floodgates come down. This would account for the way dying people may remember the whole of their past life in an instant with intense clarity, detail, and understanding."[7] Terminal lucidity in general may be best accounted for by the soul beginning, near death, to break free from the constraints of a failing brain.[8]

Why the Human Soul Cannot Simply Decompose

We have seen from neuroscience that the immaterial aspect of the human soul is a unity. It has no parts, so it cannot be split or multiplied. When the brain is split in half or largely absent or shut down, the human person is still present in whole, not in part. Even when two souls must partly share a body, as with conjoined twins, they remain whole and distinct. Thus the human soul cannot just de*compose*, as the dead human body does.

Kreeft puts it like this: The soul is not "composed" in the first place. "Whatever is composed (of parts) can be decomposed: a molecule into atoms, a cell into molecules, an organ into cells, a body into organs, a person into body and soul. But the soul is not composed, therefore not decomposable. It could die only by being annihilated as a whole. But this would be contrary to a basic law of the universe: that nothing simply and absolutely vanishes, just as nothing simply pops into existence with no cause."[9]

In general, traditional Christian and other nonmaterialist philosophers have seen the soul in these terms. The genuine evidence from neuroscience, as we have seen, supports their approach.

What Can Neuroscience Tell Us About the Immortality of the Soul?

The work of Roger Sperry, Justine Sergent, and Yaïr Pinto (chapter 1) shows that the capacity for abstract thought is not split in the way that physical perception can be split when the connections between the cerebral hemispheres are cut. The fact that abstract thought has no physical place and cannot be split is consistent with the human soul having some immaterial powers. The death of the brain does not entail the death of the soul, any more than numbers disappear from mathematics when a frustrated student flings a difficult math assignment into the fireplace.

Neuroscience points to immortality in other ways as well. When neurosurgeon Wilder Penfield performed brain operations on awake patients to remove the damaged parts that were causing seizures, he discovered that he could evoke experiences in the patient by stimulating the brain. He could evoke movements, sensations, memories, and emotions. But he could not evoke what he called the "mind," which corresponds to the capacity for abstract thought and free will (or, in philosophical terms, the immaterial spiritual powers of the soul). Penfield thought this exclusion of a whole class of thoughts—abstract conceptual thought and free will—was remarkable, and it is.

The fact that neither epileptic seizures nor operative stimulation of the brain can evoke abstract thought or free will from brain activity implies that abstract thought does not arise from the brain. Of course, normal brain activity enables us to think abstractly, if all we mean is that it makes such activity possible. We can't do calculus during a seizure or in a deep coma.

Brain processes enable your brain to work in the same way that electricity enables your laptop to work. But normal brain activity does not *generate* abstract thought. And your laptop does not *produce* the difficult letter you are perhaps typing. The mind appears to have an immaterial source, and, again, we have no reason to think that such a source is mortal by nature.

The work of Adrian Owen and colleagues on the thought processes of people in the deepest level of coma (persistent vegetative state, or PVS) shows that these individuals are not simply live bodies without minds, as was once widely believed. But that research also showed something else too: Some people whose brain damage was so devastating that they were considered effectively mentally dead were found to think on a relatively high level. Clearly, their minds were not simply what their brains were doing—their brains were nearly destroyed. That likewise points to the immateriality and thus the immortality of some functions of the soul. Although such findings may conflict with some of the beliefs of modern atheists, they would hardly have surprised ancient philosophers like Aristotle.

The near-death experience, which tends to get the most attention when we discuss the soul, is hardly a new phenomenon. It is a subset of what Harvard religion scholar Carol Zaleski calls "otherworld journeys,"[10] of which there have been many accounts in cultures around the world for thousands of years. All that's changed is that modern medical resuscitation techniques have enabled us to observe and study some aspects of these experiences in new, previously unimagined ways.

We cannot know what happens to the people who have NDEs but are not resuscitated. But there is no compelling reason to believe that they have suddenly ceased to exist, and there are good reasons to suppose otherwise. That's probably part of the reason why almost all human beings have belonged to one or another tradition that accepts the immortality of the soul.

What Can We Learn About the Human Soul from Everyday Life?

Our souls act on our bodies in a number of normal ways that researchers have been able to study. An everyday example is the *placebo effect* in medicine. Hundreds of studies have shown that many participants start to get better when they are given a sugar pill, provided that it is administered as if it were a treatment. It turns out that if doctors can get patients to believe that help is on the way, that belief has a positive physical effect on the body. The effect is even growing, according to a recent paper, perhaps because, from long experience, health care researchers are better at stimulating it![11] This has nothing to do with the body as such; it is caused by the abstract belief, based on reasonable inferences, that help is at hand.

Throughout recorded history, people of many faith traditions have also reported encountering a heightened reality through meditation. Neuroscience provides new tools for examining the claim. For example, Buddhist monks in the Tibetan tradition can apparently alter their metabolisms through prolonged meditation.[12] These monks, it must be stressed, devote their entire lives to meditation and asceticism.[13] It is not something that just anyone can decide to do. But the fact that it

Immortality of the Soul Is a Reasonable Belief

can be done at all is an example of abstract beliefs having a material effect on the body.

Similarly, neuroscientists Mario Beauregard and Vincent Paquette studied Carmelite nuns engaged in contemplation (a preferred Christian term for meditation) at the University of Montreal. Collectively, the fifteen nuns had spent roughly 210,000 hours in prayer. Beauregard and Paquette found that, when scanned during contemplation, some showed a marked alteration of brain activity associated with consciousness:

> When the nuns were recalling autobiographical memories, the brain activity was different from that of the mystical state. So we know for certain that the mystical state is something other than an emotional state. The abundance of theta activity during the mystical condition clearly demonstrated a marked alteration of consciousness in the nuns. It is noteworthy that previous QEEG studies have shown increased theta activity in the frontal cortex during a type of Zen meditation called Su-soku, and a blissful state in meditation (Sahaja Yoga meditation).[14]

Even among far less committed (but nonetheless serious) meditators, prolonged meditation has been found to alter the brain, with the changes detected mainly in the frontal and parietal lobes.[15] Researcher and radiologist Andrew Newberg and his colleagues stress that the participants' abstract and immaterial ideas genuinely affect their brains and bodies:

> We do not believe that genuine mystical experiences can be explained away as the results of epileptic

hallucinations or, for that matter, as the product of other spontaneous hallucinatory states triggered by drugs, illness, physical exhaustion, emotional stress, or sensory deprivation. Hallucinations, no matter what their source, are simply not capable of providing the mind with an experience as convincing as that of mystical spirituality.[16]

Thus it seems that the relationship between the soul and the body is a two-way street. That's what we might expect if the human soul is a real, though immaterial, entity. The things it does make a measurable difference.

Why This Question Was Hard to Study in Recent Decades

If the relationship between the mind or soul and the body is a two-way street, does spirituality—letting the soul guide the body—change behavior over the long term in a measurable way? In the early twentieth century, researchers tended to discredit the idea that the mind impacts the body. According to Harvard Medical School's Herbert Benson (1935–2022), in the 1930s, the *Index Medicus* did not offer any references to the way mental states might affect physiology.[17] In the 1960s, he had difficulty getting his colleagues to even accept that mental stress could be a factor in high blood pressure.[18] Yet he noted that when he encouraged his own patients to relax so as to avoid disrupting healing processes, four out of five chose prayers from their religious tradition, whether it was Christian, Jewish, Hindu, or Buddhist.[19]

Benson devoted his career to the study of how mental phenomena, including meditation, affect physical health. Over the years, he published a number of influential papers and books on the consistent relationship between spiritual practices and physical health, including *Timeless Healing: The Power and Biology of Belief* (1996).

The research climate has gradually begun to become more open-minded about the mind's influence on the body. Epidemiologist and psychiatrist David B. Larson (1947–2002) observed in 1995:

> What is perhaps most surprising about these negative opinions of religion's effect on mental health is the startling absence of empirical evidence to support these views. Indeed, the same scientists who were trained to accept or reject a hypothesis based on hard data seem to rely solely on their own opinions and biases when assessing the effect of religion on health.[20]

Working with Jeff Levin and Harold Koenig, Larson provided a systematic review of the available literature.[21] One outcome was *The Faith Factor* (1993),[22] in which he, Dale Matthews, and Constance Barry reviewed 158 medical studies on the effects of religion on health. Of the studies, 77 percent showed a positive clinical effect.[23] Of course, because the mind's influence is real, it can work both ways. Medical patients who report a negative relationship with their religion have been found in some research to be more likely to die.[24]

Overall, however, spirituality is usually also associated with better psychological welfare. A Baylor University study reported, "People who consider themselves more accountable to a god report higher levels of three of the four variables of psychological wellbeing. The association was stronger in people who pray more often, suggesting accountability accompanied with prayer enhances psychological wellbeing for believers."[25] A 2021 study from the University of South Florida found that "spiritual people tend to experience thought patterns that are more organized and provide deeper sense of meaning. This meaning can help be a sort of mental anchor during times of distress."[26]

Indeed. Anchors work because they are anchored to something solid that we can't usually see.

Spirituality May Influence Longevity

A prospective study involving almost four thousand seniors (ages 64–101), conducted from 1986 to 1992, found that meditation, prayer, and Bible study were associated with greater longevity.[27] Similarly, a more recent study found higher life expectancies for older Black American men who attend religious services than for others.[28] Attending religious services does not equate to spirituality, of course, but the two factors certainly overlap.

Newberg, who has studied spirituality and health for thirty years, urges us to interpret such evidence from a holistic perspective: Many of the physical and mental health benefits of spirituality are due to the lifestyle it encourages—for example, avoiding excess drink or drugs.[29] He's likely right, but choosing to follow a lifestyle consistent with spiritual beliefs is the abstract decision of a human soul capable of making a choice.

A better health outcome does not depend on the specific religion followed.[30] That isn't surprising if the good health outcome stems from the immortal human soul's effect on the brain and body. The driving force is the reality of the soul at one end of a two-way street, making wiser choices. Ultimate reality, in a theological sense, inhabits a higher plane.

Near-Death Experiences Do Not Establish Ultimate Truths

All human consciousness, even the most abstract or mystical, is conditioned by what our minds *can* understand. For example, a traditional assumption has been that no one can see God and live (except insofar as God takes steps to permit it). Thus, in a famous passage in the Old Testament, the prophet Isaiah says, "'Woe to me!' I cried. 'I am ruined! For I am a man of unclean lips, and I live among a people of unclean lips, and my eyes have seen the King, the LORD Almighty'" (Isaiah 6:5). Paul the apostle, who also had mystical experiences,[31] reminded hearers, "However, as it is written: 'What no eye has seen, what no ear has heard, and what no human mind has conceived'—the things God has prepared for those who love him—" (1 Corinthians 2:9). We should not *expect* to easily understand ultimate reality.

In the same way, we should not be surprised that many South Asians report meeting the Hindu king of the dead during a near-death experience, while Americans more often claim to have met Jesus.[32] Generally, we identify what we can recognize. This is also true in the everyday world around us. For example, we see the colors of the spectrum of light in a way that is made possible by our human eyes. But many animals see colors we don't because their eyes are engineered differently.[33] Perhaps near-death experiencers begin by seeing the unseen world in terms we can understand.

Both Gary Habermas and Michael Sabom, near-death researchers, caution that based on their research, NDEs cannot be used by themselves to establish a specific theology. As Sabom puts the matter in *Light and Death*, they are not "road signs pointing to a person's ultimate destiny."[34] They are simply witness reports of what people saw along the way.

How Can Minds See Without Eyes?

Kenneth Ring, who has studied near-death experiences of the blind, found that some individuals who were blind from birth reported NDEs that were highly visual, with content consistent with the NDEs of sighted persons. In some cases, the evidence was corroborated.[35] That implies that a form of sight is the usual means by which the mind apprehends information, though not the only possible one.

It all adds up to this: We are destined to be immortal whether we like it or not. We must choose what that means for our futures. That brings us to the basic reality and critical importance of free will.

CHAPTER 8

Free Will Is a Real and Intrinsic Part of the Soul

YEARS AGO, I HAD a patient with an alien hand. She had had a small stroke in her corpus callosum, that huge bundle of nerve fibers that connects the right and left halves of the brain. Sometimes damage to the back part of the bundle of nerves disconnects activities of the two brain hemispheres a bit, so that one side of the body seems to act on its own. When my patient was sitting quietly, her left hand would move on its own, rising up and touching things without her control. She found it creepy, as you might imagine. Fortunately it stopped happening after a couple of years, as alien hand syndrome usually does.

Does this mean that we really don't have free will? After all, if one part of our body can, after a slight disconnection in the brain, act on its own accord, without our willing it, is it possible that everything we do is without our willing it?

Free will denial is probably the majority view among neuroscientists and very popular among scientists generally. For them, alien hand syndrome is just a quirky manifestation of a deep truth about us—we are not free to choose. We are utterly controlled by our brains, in the same way that a chemical reaction or an electrical current is controlled by physical laws. For example, in 2023 eminent Stanford professor of neurology Robert Sapolsky, author of *Determined*, announced after forty years of study of humans and primate apes, "We are nothing more or less than the sum of that which we could not control—our biology, our environments, their interactions."[1] Sam Harris, a well-known American atheist and neuroscientist, puts it like this: "Free will is an illusion. Our wills are simply not of our own making. Thoughts and intentions emerge from background causes of which we are unaware and over which we exert no conscious control."[2]

Philosopher Massimo Pigliucci,[3] evolutionary biologist Jerry Coyne,[4] astronomer Avi Loeb,[5] neuroscientist David Eagleman,[6] theoretical physicist Sabine Hossenfelder,[7] and physicist Larry Krauss,[8] to name just a handful of prominent science figures in media, would agree, though each puts an individual stamp on the argument. Historian Yuval Noah Harari even claims that the idea of free will is dangerous: "If governments and corporations succeed in hacking the human animal, the easiest people to manipulate will be those who believe in free will."[9] With so much basic agreement among high science influencers, it is no wonder that the idea often filters down into popular culture that free will has been refuted by science. As we will see, it has not.

Does Denial of Free Will Even Make Sense?

In any event, it is not clear that their view that free will is an illusion makes any sense. Imagine a patient walking into a doctor's office and telling the doctor about a belief that he can't get out of his head: "I believe, Doctor, that I'm controlled by an alien force and have no control over anything I do. My thoughts and feelings and movements are not my own—they are forced upon me. I'm a robot made of meat. What can I do about this?"

The good doctor would no doubt be concerned for the patient's sanity and recommend a psychiatric evaluation.

On the other hand, if the patient added, "And I'm a neuroscientist who does brain research," we can imagine the psychiatrist exclaiming, "Well, you're in the mainstream of scientific opinion in the best universities and laboratories. Please, come and lecture to my medical students about cutting-edge neuroscience!"

Anyone who really believes that they are a meat robot with no agency and no control over their thoughts would seem to be in need of psychiatric care. Without the veneer of modern science, denial of free will would be recognized as a mental illness.

There are good reasons to affirm free will. Here are four of them: (1) the fact that we all behave as if free will is real, (2) evidence from logic, (3) evidence from physics, and (4) evidence from neuroscience.

1. Human Beings Act as Though Free Will Is Real

Consider the following scenario: You walk into a room in which a neuroscientist is busy typing away on his laptop, writing an article for a science journal in which he denies free will. You deliberately tip your hot mocha latte over onto his laptop, ruining his computer. The free will denier will scream in outrage, of course, "Why did you do that? You just ruined my computer!" You reply, "I had no choice. Just like you say in your article, I have no free will and thereby no capacity to choose. My brain chemicals made me do it. I'm no more morally responsible for destroying your laptop than my coffee cup is. You can't blame me!"

Will the free will denier accept your excuse that you couldn't have done otherwise? Of course not. You're going to be paying for a new laptop and he might choose to sue you and/or have you charged. Everyone actually believes in free will, even materialists who say they don't. Free will denial is an ideological fashion, not a rationally consistent belief.

Even free will deniers generally hold themselves and others to objective moral standards (don't steal, don't murder, don't lie, don't rape, and so on). Yet if free will weren't real, it wouldn't be fair to hold any of us to moral standards that we could not choose—we don't hold cattle or sea slugs or machines to moral standards. Human society is predicated on the reality of free will—our legal system, our educational system, our economic system, and nearly every aspect of our culture and our personal lives only make sense if we have the genuine ability to choose. Nothing in normal human thought and action makes sense without free will.

Of course, it is possible that despite the way we naturally act, there really is no free will. But the belief that there is no free will does not typically cause people to be in closer touch with reality. David Berkowitz, the infamous Son of Sam killer, claimed that he was controlled by his neighbor's dog.[10] But that is a hallmark not of a normal human relationship to reality but of mental illness.

Second, belief in free will plays an important part in human achievement as well. If free will isn't real, we can't give ourselves or others credit for initiative or for accomplishments—neither reward nor blame can be justified if we have no choices.

Societies whose inhabitants don't believe in free will tend, unsurprisingly, to drift toward totalitarianism. As political philosopher Hannah Arendt (1906–1975) has pointed out, totalitarian societies view human behavior as driven by economic and racial dynamics that are largely out of the control of individuals. Humans are thought to move as masses, as human livestock, bereft of rational choice. Denial of free will is a cornerstone of totalitarian government.[11]

Celebrity skeptic Carl Sagan was fond of saying that extraordinary claims require extraordinary evidence.[12] Do those who deny free will really have enough evidence of sufficiently high quality that we should set aside all human experience? They are far from meeting any such standard, as we shall see.

2. Denial of Free Will Involves Self-Refuting Illogical Statements

Consider the proposition "Free will is an illusion because all nature, including our actions, is completely determined by the laws of physics." As a logical argument, that statement must be either true or false.

If it were true, then the very statement "Free will is an illusion..." is generated wholly by our brain chemicals and brain processes, not by reason or contemplation of truth. Our brain processes would be caused solely by the laws of physics. We don't look at a splash of spilled ink on the floor and say, "I agree with what it is telling me." It is not telling anyone anything. We only pay attention to opinions generated by human beings who have the capacity to choose to tell the truth. That means that there is such a thing as choosing the truth. Free will denial is self-refuting by nature.

We can see how that works by a simple illustration. Suppose you tip over a cup full of Scrabble pieces and the letters fall to the floor in a pattern that reads, "I want to move to the moon." The arrangement of letters from the accidental spill is not *un*caused—it's completely caused by the laws of physics. Gravity, friction, and air resistance combined to wholly determine this combination of letters. And yet the fact that you are unlikely to literally want to suddenly move to the moon will probably be taken for granted by every witness.

The point is that the laws of physics that determine the way the Scrabble pieces fall do not have any truth value. So if our thoughts are determined wholly by the laws of physics and not by our free will, then our thoughts have no truth value. Thus, thinking "Free will doesn't exist" isn't a truth claim any more than a bunch of Scrabble pieces that fall to the floor are a truth claim. To make a claim about truth, we have to have free will that is not merely determined by the laws of physics. So what the free will denier is really saying is that every idea, including his own denial of free will, is analogous to an accidentally spilled cup of Scrabble letters—a tale signifying nothing. If we lack free will and our thoughts are determined wholly by the laws of physics and chemistry, then we can't make any valid argument. Free will denial is logically self-refuting.

3. The Evidence from Physics Supports Free Will

Do the laws of physics absolutely determine everything that's going to happen? We might think so. Laws of physics—which are expressed as mathematical equations—do not have a mathematical term that accommodates free choice. Thus Albert Einstein (1879–1955) argued that "God does not play dice" with the natural world. He meant that a better theory would show that all of nature is determined like clockwork and there is no room for free will. Indeed, he told the Spinoza Society in 1932, "Human beings, in their thinking, feeling and acting are not free agents but are as causally bound as the stars in their motion."[13]

Einstein and Danish quantum physicist Niels Bohr (1885–1962) famously debated such questions in the early years of quantum mechanics, which serves as the basis for computers and so much of our technological society today. Many scientists agreed with Bohr, who argued forcefully that probability represents nature better than Einstein's certainty. At one point, Bohr ordered Einstein, "Stop telling God what to do with his dice."[14]

The key issue was, what underlies reality? Is the course of events rigidly fixed beforehand by deterministic laws of physics? In 1964, Irish physicist John Stuart Bell (1928–1990) showed that experiments can determine whether events are rigidly fixed by physical laws before they occur. Since the early 1970s, many teams of investigators have done the experiments Bell developed and have all gotten the same results— events are *not* rigidly fixed by physical laws before they occur. There is, in reality, an indeterminacy in the moment-to-moment flow of events in the world.

In 2022, physicists Alain Aspect, John F. Clauser, and Anton Zeilinger were awarded the Nobel Prize in Physics for experiments that established that, contrary to what Einstein had hoped, at the most fundamental level, nature is governed by probabilities and not by certainties. Human choices are not rigidly bound up in the laws of physics. Modern physics does not support the deterministic view of the universe held by many materialists. Rather, it has shown us that the deterministic perspective by which materialists deny free will is wrong.

The remarkable discovery of this Nobel Prize–winning research is that physical processes are not locked in by any law of physics, even a law that we have not yet discovered. The physical world is naturally indeterministic—things happen that are not dependent on any actual or possible physical law.

Since natural processes are not, and cannot be, determined absolutely by the laws of physics, there is room for human free will to act.

It is noteworthy that the philosophical implications of this groundbreaking research are a raging contemporary debate in physics and philosophy. There are several non-mainstream interpretations of quantum mechanics that do permit determinism—for example, superdeterminism, Bohmian pilot wave theory, and Everett's many-worlds interpretation. None of these theories is mainstream, however.

We think it's fair to say that Bell's and Aspect's teams' work, understood via Bohr's mainstream Copenhagen interpretation of quantum mechanics, leaves abundant room for free will.

4. *The Neuroscience of Free Will*

A fascinating set of experiments by Montreal neurosurgeon Wilder Penfield in the mid-twentieth century clearly points to the reality of free will. As recounted in earlier chapters, Penfield specialized in the treatment of epilepsy. He operated on more than eleven hundred patients who were locally anesthetized but awake during the time that their brains were exposed and stimulated. Awake surgery allowed him to pinpoint the origin of the seizures so as to remove nonfunctioning parts without damaging functioning ones.

During these lengthy operations, he took many measurements in an effort to understand the functioning of the brain and the relation between the brain and the mind. One thing he investigated was the nature of free will. His experimental technique was quite clever. He was able to stimulate movement of an arm or a leg by applying an electrode to the surface of the region of the brain responsible for the movement. When he touched the electrode to the surface of the cortex in the brain's arm control area, the patient's arm on the other side of the body would move. During the surgery, he also asked the patient to move his arm himself at times, freely, when the patient wished.

The thing to see here is that the patient could neither see nor feel whether Penfield was forcing his arm to move. After the arm moved, Penfield would ask, "Did I move your arm or did you will to move it yourself?"

> When I have caused a conscious patient to move his hand by applying an electrode to the motor cortex of one hemisphere, I have often asked him about it. Invariably his response was: "I didn't do that. You did." When I caused him to vocalize, he said: "I didn't make that sound. You pulled it out of me."... There is no place in the cerebral cortex where electrical stimulation will cause the patient to believe or decide.[15]

For over eleven hundred patients, Penfield never once encountered a situation where he stimulated a patient's brain and the patient believed that the mental and physical activity he stimulated had been freely willed by the patient himself. That is, Penfield was never able to find a "will" center in the brain that, when stimulated, evoked a patient's sense of will. Penfield inferred that this meant that the will does not come from the brain, but is a power of the immaterial mind, and by its immateriality, the will is free. Penfield rightly interpreted this finding as strongly supporting the inference that free will is real and not wholly determined by brain activity. The will seems to have a separate existence, independent of the brain. That convinced Penfield that free will is real.

Benjamin Libet and the Discovery of "Free Won't"

Arguably, the most interesting sustained neuroscience research on free will was carried out in the early 1980s by neurophysiologist Benjamin Libet (1916–2007). Libet was fascinated by the *timing* relationship between electrical activity in the brain and thoughts. He wanted to know, what exactly is going on in the brain at the moment we make a decision?

To explore this question, Libet asked neuroscience study volunteers to sit at desks. He placed electrodes on their scalps to measure brain wave activity in the cerebral cortex. He provided each volunteer with a button to push. The button recorded the exact time it was pressed, down to the millisecond. On each desk was a clock with a sweep hand that allowed the volunteer to time with reasonable precision (to about 20 milliseconds) the moment of becoming aware of a thought. Libet asked the volunteers to decide to push the button—that is, to exercise their free will—at any moment they chose and to note the exact moment they made the decision. By doing this, he could determine the precise timing between the decision to push the button and their brain wave activity.

He found that when the volunteers decided to push the button, the conscious decision was invariably preceded by a spike in brain wave activity about half a second beforehand. He called this spike a "readiness potential." At first it seemed as if what appeared to be an act of free will—a conscious decision to push a button—was in reality driven by an electrochemical process in the brain about half a second earlier, of which the volunteer was not even aware. Thus it was widely assumed that the volunteers were compelled to consciously decide to push a button by an unconscious spike of brain activity.[16] This result seemed to support a deterministic model of decision making and to refute the idea of free will. A 2021 review of responses to the findings found it "difficult to overstate" the effect Libet's experiments had on thinking about free will at the time.[17] British philosopher Julian Baggini captured the jubilant mood when he described neuroscience studies such as these as "the final nail in free will's coffin."[18]

But Libet disputed that interpretation of his work. He carried the experiment one step further. He asked the volunteers to veto the decision to push the button. That is, they were to decide to push the button and then immediately decide not to. He recorded their brain activity during this process as well. Remarkably, he found that while the decision to push the button was preceded by a readiness potential, *there was no new brain activity associated with the veto.*[19] That is, while the "temptation" to push the button did indeed appear to arise from the brain, the choice to accept or veto the decision did not. The decision appeared, in that sense, to be free from materialistic determination. He famously quipped that he couldn't say for sure that he had proved the reality of free will, but he could say that he had proved the reality of free *won't*.[20]

Libet was philosophically astute. He pointed out that his experimental understanding of choice and free will implied that human beings are bombarded with unconscious motives that are to some extent material and deterministic (that is, generated by the brain). But, he thought, we retain the indeterministic ability to freely accept or reject them. He added that his results correspond nicely with traditional religious understandings of temptation, sin, and free will.[21]

It turned out much later (2021) that the brain activity that precedes awareness of a choice, recorded by Libet and other investigators, likely represents nonspecific "noise" in neuronal networks. The noise appears to be associated with the mental state that leads to making a choice. More detailed research shows that the brain activity that is most highly correlated with actually making a decision happens simultaneously with awareness of the choice.[22]

Another recent research finding (2023) focused on the relative importance of a decision. After all, pushing a button in Libet's lab changed no one's life. What about decisions with true consequences? The researchers found something very interesting: "Meaningless choices were preceded by a readiness potential, just as in previous experiments. Meaningful choices were not, however. When we care about a decision and its outcome, our brain appears to behave differently than when a decision is arbitrary."[23] If anything, such research findings support free will. This recent research suggests that meaningful decisions—the kinds of decisions that we ordinarily most closely associate with free will—aren't preceded by characteristic brain activity such as the readiness potential and thus seem to have a nonmaterial cause—that is, immaterial free will. Given that free will matters most to us when we are making important decisions, it is surely significant that the important decisions did not seem to have a material correlate.

As things stand, Cristi L. S. Cooper perceptively notes in *Minding the Brain* (2023) that "although the vast majority of neuroscientists believe in a deterministic view of free will, many of them do not believe that Libet's experiment can be shown to do away with free will."[24] Unfortunately, the impression left by many Psychology 101 classes, that Libet disproved free will, has far more to do with the instructors' private beliefs than with the state of neuroscience research.

The Wider Implications of Free Will Denial: No Guilt Means No Innocence

Common human experience, along with logic and evidence from both physics and neuroscience, clearly point to the reality of free will. This evidence is of vital importance not only for science and psychology but also for our legal system. If there is no such thing as free will, then there is no such thing as crime, liability, or moral transgression. Neither a murderer, nor a negligent school board, nor a corrupt politician could have chosen to act otherwise. "My brain made me do it" becomes a "get out of jail free" card.

Philosophers and scientists who argue that free will is an illusion generally reply to this objection by saying that even if free will doesn't exist, penalties deter people from committing crime in much the same way as electric fences deter cattle from wandering away. Here's the problem with their view: Because we don't hold cattle morally responsible for their transgressions, we simply deny them freedom and herd them. Thus the materialist viewpoint offers no higher objective than livestock management for humans.

An even more disturbing outcome follows: If there is no guilt, there is no innocence either. Why not incarcerate people *before* they have committed crimes, as in the 2002 film *Minority Report*? The inevitable protest—"But I'm innocent!"—is meaningless under the circumstances. Without free will, we are never either guilty or innocent. It is the proposed *future* risk that the authorities seek to prevent. Because the denial of free will is part of the formula for a totalitarian state, the Soviet Union, for example, "medicalized" political dissent by forcibly hospitalizing political dissenters as psychiatric patients.[25] The pretext was that they dissented from government policy because they were unable to reason correctly. The controlling underlying assumption was that they could not have made a rational free choice to disagree with the regime.

Is Free Will Making a Comeback in Philosophy?

Overall, denial of free will doesn't make sense of human nature. To insist that our neurotransmitters completely control our choices is no different from insisting that our iPhones control our thoughts. Concentrations of neurotransmitters aren't propositions and have no truth values. Dopamine isn't "truer" than serotonin. As we have seen, there is much more to our thoughts and our choices than that.

Thus, while many thinkers dismiss free will, many others defend it. Theoretical physicist George Ellis, considered a world leader in relativity and cosmology, contends that physics has made huge strides but has not upset free will. For one thing, he argues, determinism would require all the books ever written to have been encoded in the Big Bang, an unlikely scenario.[26]

Prominent science writer John Horgan has said, "I can live without God, but I need free will." At *Scientific American*, he interviewed philosopher and free will defender Christian List, who told him that "agency, choice, and control are emergent, higher-level phenomena, like cognition in psychology and institutions in economics. They 'supervene' on physical phenomena, as philosophers say, but are not reducible to them."[27] Horgan also sensibly observes that, given how little we know about human consciousness, why *would* we just rule out free will? "Philosophers speak of an 'explanatory gap' between physical theories about consciousness and consciousness itself. First of all, the gap is so vast that you might call it a chasm."[28]

And yet, curiously, during the debate around Robert Sapolsky's new book *Determined* in 2023, which argued against free will, Horgan found himself doubting free will because it interferes with his passionate attachment to Darwinian evolution.[29] Sapolsky is a neuroscientist and a Darwinist who vigorously denies free will and embraces a deterministic and Darwinian view of nature. It's noteworthy that Horgan is willing to jettison belief in free will—which he has so staunchly and eloquently defended for years—to salvage his belief in...Darwinism.

Linguist and philosopher Noam Chomsky raises an important point here. The inability of a philosophical system to account for free will may say more about that system than about free will: "If it's something we know to be true and we don't have any explanation for it, well, too bad for any explanatory possibilities."[30] Perhaps that's why a number of younger thinkers are also beginning to treat free will as real, according to science writer George Musser, author of *Spooky Action at a Distance* (2015).[31]

Even prominent atheist thinkers like Daniel Dennett (1942–2024),[32] Richard Dawkins,[33] and Steven Pinker[34] now seem to hedge, hoping that some acknowledgment of free will may be compatible with their overall determinism. In 2023, Trinity College neurologist Kevin Mitchell mounted a defense of free will, *Free Agents*, in which he argues that evolution has somehow given us free will "without the need for any mystical or supernatural forces at play."[35] That might seem like good news for John Horgan. But of course, it's unclear how impersonal acts of nature, which has no free will itself, could give humans free will. The main thing to see here is that many people are looking for reasons to believe in free will that don't involve questioning their basic materialist assumptions.

However, their recent uncertainties don't have the cultural reach (or siren lure) of simple free will denial. That damage will take a long time to undo. The real difficulty, of course, is that free will is only really comprehensible if it is seen as one of the immaterial powers of the human soul.

Free Will Is a Real and Intrinsic Part of the Soul

We Have Libertarian Free Will

The simplest view on free will—the view that is consistent with logic, human experience, physics, and modern neuroscience—is that free will is real. We have the ability to choose our actions. But it's fair to ask: How does libertarian free will work?

From the beginning of civilization, mankind's best thinkers have contemplated the nature of free will. Plato's analogy of the chariot is helpful. Plato (c. 428–c. 348 BC), Aristotle's teacher, envisioned the human person as a chariot with a driver, pulled by two horses. The driver is the intellect, which uses reason and judgment to steer. The horses are our appetites, which are the animalistic urges and desires that pull us each day—passions for food and sex, comfort and power. The reins are the will—they are the link between the intellect and the lower passions and the means by which our intellect guides us and steers our lower passions toward destinations chosen by reasoned deliberation.

Thomas Aquinas had a similar but more nuanced view of the will, and he confirmed its liberty: "But man is freer than all the animals, on account of his free-will, with which he is endowed above all other animals."[36] The will and the intellect are immaterial powers of the spiritual human soul. The will is not determined by matter. In fact, it *cannot* be determined by matter, because it is spiritual, not material. The will can move matter as a final cause, a purpose, just as the intellect moves matter as a formal cause, an idea. The natural goal of the human intellect is the pursuit of truth, and the natural goal of the human will is the pursuit of the good. Note that the traditional account of free will as one of the powers of the human soul corresponds very well to human experience, to logic, to physics, and to modern neuroscience.

The ubiquity of belief in free will, the logic that points to the existence of free will, and the evidence in physics that there is room for free will in human action all point strongly to the reality of free will.

I believe that neuroscience evidence for free will is very strong as well. As Penfield discovered, there is no area that is the "will center." That is, our ultimate decisions to act do not appear to come from the brain and thus do not appear to be the product of mere material processes—electrical signals and neurochemicals in the brain.

As a neurosurgeon, I must point out that even a patient's consent to undergo surgery depends on the reality of free will. Consider that if free will were not real, and if our actions were wholly determined by brain impulses and chemicals, how could we consider a patient's consent to undergo surgery to be valid? After all, if free will is not real, the patient didn't really "choose" to have surgery, any more than the patient chooses to jerk his leg when the doctor tests his reflexes by tapping his knee with a reflex hammer. Countless social interactions and even the reality of human dignity depend on the presumption that we are genuinely free to make choices.

We have the free capacity—the spiritual capacity—to choose good or evil. Multiple lines of evidence point to the fact that, ultimately, we are not mere meat machines. We are free to make choices in our lives.

CHAPTER 9

Models of the Mind—Which One Fits Best?

SOME YEARS AGO, I attended a lecture at my university given by philosopher Patricia Churchland. Dr. Churchland is a prominent advocate of *eliminative materialism*. That is the view that we don't really have minds at all, just brains. Our belief that minds exist is a form of "folk psychology," and we need to get over it.

The lecture hall was packed—Dr. Churchland is highly esteemed in her field—and many audience members were neuroscientists. As we walked out afterward, a neuroscientist friend of mine—an excellent scientist but a bit of a curmudgeon—said, "I don't think she's right about there not being a mind."

I replied, "I agree. So, what is the mind? How does the brain cause the mind?"

He replied, "It's simple. The mind is caused by the cerebral cortex. The neurons make the mind. That's all there is to it. I don't know why people don't get this."

"*How* do the neurons make the mind?" I asked.

He waved me off and walked away, unwilling to be bothered by an annoying question.

Actually, the question "How does the brain make the mind?" is a central one of modern philosophy. Libraries are full of books on this question. Answers from a materialist perspective abound, but the sheer variety of answers implies that the question remains open. Behaviorism, mind–brain identity theory, and computer functionalism have all been tried. Many of us have taken courses in psychology, neuroscience, counseling, and so on in which they were or are much-discussed staples. We may have heard much less about why they don't work.

With the predictable eclipse of each of these odd theories (materialist theories of mind last a generation or two, until people figure them out), this new theory championed by Dr. Churchland, eliminative materialism, is gaining traction. Eliminative materialism is at least economical—materialism can't explain the mind, so the mind can't exist. But why deny the mind, rather than deny materialism? The mind, after all, is hard to explain away. It's hard to make a case for a belief that there are no beliefs.

The Central Question of Modern Philosophy

Let's look a little more closely at the theories that have dominated neuroscience for over a century and see how they fare against the central question of modern philosophy and neuroscience: What, exactly, is the mind?

In June 1998, Allen Institute neuroscientist Christof Koch made a bet with dualist philosopher David Chalmers. Koch picked his adversary with care; Chalmers is best known for coining the term "the hard problem of consciousness." But within twenty-five years, Koch declared confidently, the *material signature* of consciousness, that is, the specific brain circuit responsible, would be pinpointed in the brain.[1] Thus the world would know that the material signature *is* consciousness...Case closed.

Without knowing exactly what they were looking for, Koch and others then set out to find that material signature. If Koch won the wager, as he expected to do, in 2023 Chalmers would have to buy him a case of fine wine.

What Koch Was Up Against

Neuroscience does not really support a materialist model of the mind. As we saw when discussing immortality, an older *dualist* model works better: Some powers of our mind are tightly linked to brain matter and are mortal, and some powers are spiritual and immortal. For example, our senses and our ability to move our bodies is mortal; our intellect and free will are immortal.

Damaging an optic nerve, for example, has immediate, profound effects on vision. But while cutting the brain in half affects perception in subtle ways, as we have seen, it has no effect on the powers of reason or will or on the unified sense of self. In fact, the exercise of our intellect and will has a real independence from the brain. This is the model that neuroscience has actually shown us, even if it is not popular among neuroscientists.

One reason it is not popular is that we live in a culture where materialism (that is, "the brain is all there is") is dominant and largely undisputed. Thus, *Time* magazine announced to readers in 1995, "Utterly contrary to common sense...and to the evidence gathered from our own introspection, consciousness may be nothing more than an evanescent by-product of more mundane, wholly physical processes."[2] In 2007, well-known Harvard cognitive scientist Steve Pinker reiterated the fundamental doctrine, also in *Time*: "Consciousness does not reside in an ethereal soul that uses the brain like a PDA; consciousness is the activity of the brain."[3] Upping the stakes, prestigious philosopher Massimo Pigliucci made clear in 2019 that any suggestion that the mind is *not* simply what the brain does, without remainder, is "antiscientific."[4] While these statements do not reflect what the evidence shows, they exert a powerful hold on neuroscience. One of the most popular theories that grew out of this long-standing mindset was behaviorism.

Just Forget the Mind!

Behaviorism, which was very influential in the twentieth century, starts with the assumption that a concern for internal mental states adds nothing to psychology as a science. Developed in large part by American psychologist John Watson (1878–1958), it focuses on observing and drawing conclusions from our behavior only. For example, if we say that Sam is sad, we mean that Sam is disposed to certain behaviors, such as crying, describing his sadness to others, or avoiding "happy" activities. We look no deeper. Philosopher Gilbert Ryle (1900–1976), who viewed the mind as a mere "ghost in the machine," developed concepts consistent with this approach, which was expected to revolutionize psychology. Famous proponents have been Ivan Pavlov (1849–1936), Edward Thorndike (1874–1949), and B. F. Skinner (1904–1990).

The rise and fall of behaviorism is a long story, beyond the scope of this book. Briefly, the fact that it leaves out our inner experience of human life doomed it from the beginning. Many people today might have a hard time understanding why anyone ever took it seriously. But in the twentieth century, much of the university elite simply assumed not only that the mind is merely the physical functions of the brain, but also that everything about the mind would one day be explained in material terms, in the same way that the output of a computer can be explained by the workings of its physical parts.

Models of the Mind—Which One Fits Best?

Unexpectedly, a linguist played a key role in behaviorism's demise. MIT's Noam Chomsky began by addressing a fundamental question in linguistics: "How did human language arise?" Many animals use sounds and signals to convey information, but only humans use it with almost unlimited possibilities for communication, particularly of abstract ideas.

Until the 1950s, the prevailing view of language's origin was behaviorist. While we have no real information about the ultimate origin of human language, its development in infants seemed reasonably well understood. Babies babble and parental reinforcement gradually trains them to use words in a grammatical way. On that view, language is a learned behavior driven by stimulus and reward via trial and error, reward and reinforcement.

In the 1950s, Chomsky (a graduate student at the time) observed that trial and error could not account for the astonishing ability of young children to pick up the language spoken around them. Children quickly learn to speak, Chomsky pointed out, despite a "paucity of stimulus." That is, there was not a shred of evidence in the psychology literature that children learn language by trial and error. Language is too complex. For all languages, even very young children use grammar naturally without having been taught or encouraged to do so.

Chomsky thought that there must be an innate, conceptual "language organ" that allows us to instinctively recognize and use grammar and understand what words mean. As an example, he offered behaviorists a challenge in the form of a sentence: "Colorless green ideas sleep furiously."[5] It's a safe bet that this sentence is unique—no one was likely to have ever heard it before, so no one could know if it was grammatically correct just by experience nor just by its meaning (it's nonsense). Yet we intuitively know it's grammatically correct. No behaviorist theory can account for our ability to speak an unlimited number of meaningless grammatically correct sentences. We have an innate ability to generate grammar.

Chomsky's point was that grammar is independent of meaning and independent of experience. Just as we can construct an unlimited number of grammatically correct nonsense sentences that we've never heard before, we can also construct an unlimited number of grammatically incorrect meaningful sentences. For example, a Pete Arno cartoon character, contemplating a crashed plane, once muttered, "Well, back to the old drawing board." That sentence lacks a verb but everyone knows exactly what it means. It even became a catchphrase.

In short, the most striking characteristic of human beings—our capacity to use language—does not and cannot have a merely behavioral explanation. Language cannot be accounted for by the behaviorist theory of reinforcement and reward. Human language acquisition is wholly unlike the training of pigeons. Within a decade of Chomsky's groundbreaking work, behaviorism was swept from the field of psychology as a conceptual framework for understanding the mind, though it remains helpful in the treatment of some emotional disorders.

Identity Theory: The Mind Is the Brain

Australian philosopher David Armstrong (1926–2014), author of *A Materialist Theory of the Mind* (1968), may have been the first thinker to popularize the idea that mental states are identical with brain states. His *identity theory* does not propose that the brain causes the mind; rather, the brain *is* the mind. Thus, a particular mental state—your love of strawberry ice cream, for instance—is *the same thing as* a particular brain state, right down to a specific activation of neurons in your temporal lobe, perhaps.

On Armstrong's view, if you could open a living human brain and inspect it, you would actually be seeing that individual's thoughts as they occurred. You would need to understand how exactly the thoughts are manifested in brain tissue, which would take quite a bit of science and a lot of philosophy. But if the theory is right, it is doable.

Identity theory is an extreme example of *reductionism*, an approach to science in which natural phenomena can best be understood by reducing them to their most elementary material parts. Reductionism does not merely propose that we can understand things better by taking them apart. It proposes that everything consists merely of those parts. Identity theorists point to atomic theory as an analogy. A kitchen table is really just a collection of atoms. Aside from the atoms, they claim, there is nothing else about the table. And just as the table is its atoms, the mind is the brain. The *Internet Encyclopedia of Philosophy* tells us, "philosophers of mind divided themselves into camps over the issue."[6]

As you may imagine, the camps didn't get along. For example, identity theory proposes that a mental state—your desire to bring an umbrella because you believe it will rain—is really a brain state, a configuration of neurons in a specific brain area with a particular electrochemical state. But whereas the relationship between mental states is logical (that is, it's logical to bring an umbrella if you believe it will rain), the relationship between brain states is physical (electrochemical and so on). *And there is no known point of contact between the laws of logic and the laws of nature.* The laws of chemistry don't contain the laws of logic. If the mind is merely the brain, how could we think logically?

Identity theory has been in eclipse recently, and justifiably so. In addition to its inability to account for logical thought, it violates what has come to be called Leibniz's Law: In order to be identical, two things must be *completely* identical. But the mind and brain share no common properties at all. The mind has ideas and feelings; the brain has blood and flesh. Identity theory is not merely wrong. It's completely wrong.

Functionalism: The Mind Is a Meat Computer

Neither behaviorism nor identity theory captured the relationship between the mind and the brain. But the 1970s offered an exciting new prospect: computer *functionalism*. Neuroscientists and philosophers of mind noted that a computer is not simply a chunk of metals, silicon, and plastic, but a device that computes—it transforms an input signal to an output signal according to an algorithm. Functionalists think that the mind–brain relationship works that way.

Pain, for example, is an input that produces crying, complaining, and wincing as an output. Functionalists locate the mind in the functional relationship between cause and effect, the output. The mind (the output) is seen as a kind of computation. The mind–brain relationship then falls neatly into the paradigm of the software–hardware relationship—the brain is the hardware and the mind is the software. Philosopher Jerry Fodor (1935–2017) was its "most explicit and influential advocate."[7]

One must admit that the theory offers a certain economy and elegance, fit for the computer age, especially when computers can sometimes seem, in an eerie way, to behave like human beings—and when AI is training them to behave even more so.

An algorithm is a set of simple mechanical rules that can give rise to enormously complex outputs, which is attractive if we are looking for simple explanations for complex things. And indeed, neuroscientists commonly characterize brain activity using computational models. Transhumanists even believe that, following the computer analogy, a mind could just as easily be loaded into a computer as into a human being, provided the proper functional relationships are maintained (and there were no power failures!).

But here's the problem: Nothing about the interaction between simple mechanical steps in a computer appears to generate mental states.

Models of the Mind—Which One Fits Best?

Consider what goes on inside a pocket calculator. It calculates swiftly and precisely, far beyond the power of the unaided human intellect. But we don't suppose that the pocket calculator understands arithmetic, any more than a watch knows the time. *We* understand arithmetic, helped by a calculator. *We* know the time, helped by a watch. Computation is not intelligence. It's a tool used by human beings.

And that's a good thing too. Fortunately, your laptop doesn't have an opinion about the essay you type on it. In fact, it would be downright inconvenient if it did—you would have to buy a new laptop that agreed with each new opinion you typed. Microsoft would love it, but they're out of luck. Computers do not and cannot think. Thus our minds are not computers.

Berkeley philosopher John Searle noticed that difficulty and offered the well-known *Chinese room problem* in a 1980 journal article[8] to explain what is wrong with functionalist theory. Here is *Britannica*'s version:

> Searle's thought experiment features himself as its subject. Thus, imagine that Searle, who in fact knows nothing of the Chinese language, is sitting alone in a room. In that room are several boxes containing cards on which Chinese characters of varying complexity are printed, as well as a manual that matches strings of Chinese characters with strings that constitute appropriate responses. On one side of the room is a slot through which speakers of Chinese may insert questions or other messages in Chinese, and on the other is a slot through which Searle may issue replies. In the thought experiment, Searle, using the manual, acts as a kind of computer program, transforming one string of symbols introduced as "input" into another

string of symbols issued as "output." As Searle the author points out, even if Searle the occupant of the room becomes a good processor of messages, so that his responses always make perfect sense to Chinese speakers, he still would not understand the meanings of the characters he is manipulating. Thus, contrary to strong AI, real understanding cannot be a matter of mere symbol manipulation. Like Searle the room occupant, computers simulate intelligence but do not exhibit it.[9]

Searle asked: Is this process of mechanical calculation, which entails no personal understanding, really an adequate model of the mind? Nineteenth-century philosopher Franz Brentano (1838–1917) pointed out that one thing that absolutely distinguishes mental from physical states is *intentionality* or "aboutness." We think *about* people, places, things, and ideas. Brentano called intentionality "the hallmark of the mental." By contrast, bricks, raindrops, and electrons flowing in computers aren't "about" anything, in and of themselves.

Intentionality is precisely what computation lacks. Computation has no *meaning* itself. That is why a word processing program works just the same for composing a chapter that supports our thesis as for one that denies it. The meaning is what human beings put into it and what other human beings take out of it.

So while the brain can be modeled as a computer, in the (perhaps trivial) sense that the activity of neurons can be described as inputs and outputs according to neurochemical rules, the mind is not computation. The mind is the opposite of computation, because thoughts have meaning, to which computation is blind. Ironically, as a theory of mind, computer functionalism has one key merit: It is a concise description *of what the mind is not*.

Although functionalism was originally developed by American philosopher Hilary Putnam (1926–2016), he apparently abandoned the view in the 1990s because "similarity of function does not guarantee identity of subjective experience, and, accordingly, that functionalism fails as an analysis of mental content."[10]

Eliminative Materialism: There Is No Mind; It's All Just Folk Psychology

Dissatisfaction with behaviorism, identity theory, and the computational theory has led to the popularity of one of the dominant twenty-first-century theories of mind: eliminative materialism—the subject of Patricia Churchland's lecture at my university. *Stanford Encyclopedia of Philosophy* tells us that Patricia and her husband Paul Churchland's writings "have forced many philosophers and cognitive scientists to take eliminativism more seriously."[11] Briefly, they hold that the mind simply does not exist. It's mere "folk psychology," a "false and misleading account of the causes of human behavior."[12] Patricia Churchland in particular has argued that our value systems originate as peptides such as oxytocin and vasopressin.[13]

But the Churchlands are hardly alone. Here is a sample of currently fashionable views along the eliminationist spectrum:

- "The terms 'mind' and 'mental' are messy, harmful and distracting. We should get rid of them," says philosopher Joe Gough (2021).[14]
- University of Mainz philosopher Thomas Metzinger offers, "The first-person pronoun 'I' doesn't refer to an object like a football or a bicycle, it just points to the speaker of the current sentence. There is nothing in the brain or outside in the world, which is us. We are processes" (2017).[15]

- It is time to give up on consciousness, say neuropsychologist Peter L. Halligan and psychologist David A. Oakley: "Our proposal feels personally and emotionally unsatisfying, but we believe it provides a future framework for the investigation of the human mind—one that looks at the brain's physical machinery rather than the ghost that we've traditionally called consciousness" (2021).[16]
- Duke University philosophy professor Alex Rosenberg is more straightforward than most about what the trend means overall. Naturalism, he says, "forces us to say no in response to many questions to which most everyone hopes the answer is yes. These are the questions about purpose in nature, the meaning of life, the grounds of morality, the significance of consciousness, the character of thought, the freedom of the will, the limits of human self-understanding, and the trajectory of human history" (2020).[17]

The Snake That Eats Its Own Tail

Essentially, because no materialist (or naturalist) theories account for the relationship between the material brain and the immaterial mind, all these philosophers simply jettison the mind, rather than jettison materialism. But that means discarding even the truth value of science. For example, University of California–Irvine cognitive scientist Donald Hoffman, author of *The Case Against Reality: Why Evolution Hid the Truth from Our Eyes* (2019), explains that we evolved for survival, not for apprehending reality:

> It leads to a crazy-sounding conclusion, that we may all be gripped by a collective delusion about the nature

of the material world. If that is correct, it could have ramifications across the breadth of science—from how consciousness arises to the nature of quantum weirdness to the shape of a future "theory of everything." Reality may never seem the same again.[18]

Philosopher Edward Feser spells out the implications of all this, by way of warning:

> Since science is as laden with intentionality as anything else, you will have to eliminate the very science in the name of which you are carrying out the elimination; and since philosophy (including eliminative materialist philosophy) is also as laden with intentionality as anything else, you will also have to eliminate eliminativism. Eliminativism is a snake that eats its own tail.[19]

Well then, eliminative materialism is a very high-maintenance philosophy. Evading self-refutation is a full-time job. But what of its relationship to reality? In "The Illusion of Conscious Thought," University of Maryland philosopher Peter Carruthers stakes a middle ground: Consciousness "is not direct awareness of our inner world of thoughts and judgments but a highly inferential process that only gives us the impression of immediacy."[20]

We disagree. Consciousness is intimate to us—we have thoughts and feelings with an immediacy that is not at all "inferential." I don't infer my thought, I *think* it. I don't infer my pain, I *feel* it. The eliminative materialists' assertion that consciousness is an illusion is not even wrong—it's self-refuting. To have an illusion, one must be conscious.

Our experiences of beliefs, desires, and so on are not "theories" or "inferences" that can be proven wrong. Our mental states are real experiences, not falsifiable hypotheses.

The most obvious objection, though, remains the one that Feser raised: To believe that eliminationism is a reasonable philosophy is to deny that there are any beliefs, including its belief that there are no beliefs. That is gibberish. Eliminative materialists have replied to this seemingly unanswerable critique with a variety of rhetorical evasions that need not detain us here.

We take eliminationism seriously for two reasons: It is a popular and ambitious conceptual mess even if it is self-refuting. But, more important, as Feser also points out, all forms of materialism are actually "disguised forms of eliminative materialism."[21] They eliminate the mind using different terminology, but they always amount to eliminating the mind. They become a confession, of sorts, that materialism is powerless to explain the mind.

Materialist Blunders

Philosopher Roger Scruton (1944–2020) quipped (and we paraphrase) that modern neuroscience is a vast collection of answers with no memory of the questions.[22] Libraries are filled with elegant papers detailing the functions of neurons and brain waves and the intimate anatomy and physiology of the brain. Yet the conceptual framework in which these investigators tried to link their experimental findings to a deeper understanding is a morass of confusion and self-refutation. We need, as neuroscientist M. R. Bennett and philosopher P. M. S. Hacker have emphasized,[23] conceptual hygiene in neuroscience, a conceptual framework that is true to logic, to neuroscience, and to our everyday experience. The traditional account of the mind is much truer: The mind (several powers of the soul) is what the living person is and does, the sum of activities that make us alive.

Models of the Mind—Which One Fits Best?

Materialism's Dilemmas of Consciousness

Just as some thinkers say that nothing is conscious, others now insist that everything is conscious. Most have staked positions somewhere in between. Here are some of the dilemmas those who are looking for a material signature of consciousness in the brain find themselves addressing.[24]

What qualifies as consciousness? Dogs, cats, and chimpanzees are obviously conscious. But what about clams and corals? Are bacteria and viruses conscious? Does mere life or physicality of any sort entail consciousness? Some philosophers and scientists argue for that view, called *panpsychism*.[25]

Philosophers have proposed several characteristics of consciousness of life-forms.[26] *Sentience*—first-person experience—is one of them. Does sentience mean response to the environment? Bacteria can sense and respond to chemical gradients. But are bacteria *aware* of gradients? Are bacteria sentient? What about machines that respond to their environment—closed-circuit cameras and audio recording devices? Are machines sentient? Are machines conscious? Is sentience essential to consciousness? We don't know. And we don't know how to know.

How do we detect consciousness? *Wakefulness* is treated as synonymous with consciousness. But we experience vivid dreams and can behave as if conscious while under hypnosis. Mystics can be intensely aware of a transcendent reality even when they might be considered by observers to be unconscious. So wakefulness isn't essential to consciousness, nor does it define it.

Self-consciousness implies not only awareness, but also awareness that one is aware—an awareness of self. This qualification adds a bit of rigor to any proposed definition, but it too features serious problems. Very young infants seem unaware of themselves as distinct entities—consider, for example, a baby's fascination when he discovers his own hands for the first time at a few months of age. Even though he was sucking his thumb long before birth, it takes him a while to grasp that his hands are part of him and that most of his post-birth environment is not.

But even adults usually go about daily activities without explicit self-awareness. Is a concert pianist thinking of himself when he plays a concerto? Are you thinking of yourself as you read this book? While adults generally have the capacity to be self-conscious, self-consciousness isn't an essential component of consciousness.

We feel as if we know what consciousness is, but when we try to define it precisely, we flail. Perhaps that is because consciousness is that *by which* we think, not *that which* we think. Consciousness is the ability to think, not thinking itself. Consciousness is like contact lenses, which are invisible to us yet enable us to see. Consciousness is the means, not the object, of thought. We can't know what consciousness is because we see and know everything through it.

So Who Won the Bet?

At a conference in New York City in 2023, neuroscientist Christof Koch conceded that twenty-five-year bet to dualist philosopher David Chalmers.[27] The "signature of consciousness" had not been captured. Chalmers reassured the readers of *Nature* that "over the years, it's gradually been transmuting into, if not a 'scientific' mystery, at least one that we can get a partial grip on scientifically." Handing over the case of fine wine, Koch, with Chalmers's blessing, vowed to fight on.

But shortly thereafter, a huge uproar racked the discipline of consciousness studies. Koch is an exponent of the leading theory of consciousness, *integrated information theory*, which is a complex mathematical model of brain physiology that posits a physical basis for consciousness.[28] It was slammed by 124 prominent neuroscientists and philosophers of mind as "pseudoscience."[29] For one thing, the theory was tarred as panpsychist[30] *and*—this was probably worse in the signatories' view—it left open the possibility that unborn children might be conscious and therefore deserving of protection.[31]

At that point, it wasn't even clear how much of the current study of consciousness is even science and how much is politics. Given the logic and the evidence, there is no reason to think that materialist neuroscience is more "scientific" than the traditional dualism that it hoped to replace. The failure to find a "material center of consciousness" in the brain should be taken seriously—it's not a failure of science; it's a success of science. It's strong scientific evidence that points to the truth about the human soul. We haven't found the material center of consciousness in the brain because the human soul isn't in the brain. Consciousness has no location. The human soul is *spiritual*.

But now another question arises. The human soul has an immaterial and immortal element that cannot just be explained away, so how could it have evolved from an animal mind?

CHAPTER 10

The Human Mind Has No History

I'VE ALWAYS HAD A problem with evolution. In my youth, it wasn't a religious problem—I was an atheist—but it was a logical problem. Even in high school, I noticed the difference between Darwin's theory of evolution and other science theories. In physics we would study mechanics, electromagnetism, and atomic theory, and I was mesmerized. In chemistry, we would study molecules, reactions, and other fascinating stuff. In biology, we would study life itself—cells, hearts, and brains. I couldn't get enough of it. I'd stay after class and ask the teacher questions.

But when we came to Darwin's theory of evolution, things were different. It seemed like a trivial idea that was trying to be science. You could sum it up: "Things change, and survivors survive," dressed up in a lot of jargon like "punctuated equilibrium," "allopathic speciation," "adaptive radiation."

It got worse in college. My evolutionary biology professor was a world expert on bird evolution and, no kidding, he even looked a bit like a bird, which the students privately found amusing. He taught evolutionary biology as an argument, more than a theory—denouncing creationists and making evolution seem more like dogma than science. I was not yet a Christian, but even a thinking atheist might have a problem with a theory that is essentially a tautology: "What are survivors? Those who survive. Who survives? Those who are survivors." That's pretty weak for science.

Of course, as I came to see the mountain of evidence for design in the world, and opened my mind to God's existence, my skepticism about Darwinism became outright disdain. Darwinism is atrocious science. For example, twenty years ago, I asked a very prominent science journalist who was a Darwinian believer just how much biological information—in DNA, biomolecules, cells, organs, organ systems, and whole bodies—can a Darwinian process generate? I wanted numbers, not hand-waving. Of course, the Darwinian journalist couldn't tell me, and no one can.

Darwinism is the only scientific theory in which a scientist can make claims about the power of the theory without measuring anything. In cosmology, if you claim that the redshift is evidence for the Big Bang, you can measure the shift and see if it corresponds to what general relativity would predict. In chemistry, if you claim that a given chemical reaction will generate so many joules of energy, you can measure the temperature. In neuroscience, if you claim that a certain pattern of brain waves indicates a seizure, you can measure the brain waves during a seizure and prove (or disprove) your hypothesis. Darwinists, by contrast, propose measurement-free explanations. Recall, for example, that a Darwinian explanation for near-death experiences is the unsupported claim that humans experience something similar to what opossums experience when they are playing dead. Not only is there no possible measurement, there is no reasonable evidence for the claim nor any remote resemblance between the two types of experiences.

As we might expect, Darwinists have applied their measurement-free science to the mind. Theories abound about the evolution of the human mind. For example, we are told that being able to think abstractly helped our ancestors hunt better in groups.[1] And yet wolves and killer whales hunt very efficiently in groups without the benefit of any abstract reasoning at all. Darwinian theory teems with such stories. Darwinian theory *is* stories.

So, in this chapter, we ask and try to answer a question at the heart of biology, neuroscience, and evolutionary theory—did the human soul evolve by natural selection? Or does it have no earthly history?

The Claim That Humans Are Not Special

We often hear popular science figures heap contempt on the idea that humans are exceptional among life-forms. It has been dismissed by evolutionary biologist Stephen Jay Gould (1941–2002) as "cosmic arrogance."[2] And yet no one has any idea how human consciousness came to exist.

For example, do dogs ask, "Are we exceptional?" Do trees wonder about anything at all? The very *existence* of such a question shows that human consciousness is exceptional. We humans are the only living creatures on this planet who think about mathematics, science, literature, philosophy, religion, and the whole range of abstract ideas. Human exceptionalism is obvious and irrefutable. So *why* is such overwhelming evidence ignored or derided?

Perhaps the reason is that the specifically human quality—a capacity for abstract thought—that leads to the question creates serious problems for materialism and for Darwin's famous and widely accepted theory of evolution. For example, prominent psychiatrist Ralph Lewis, who tries to find an evolutionary origin for specifically

The Human Mind Has No History

human qualities, has said that consciousness is "very much a biological phenomenon."[3] We are asked to believe that the human mind evolved by a long, slow process of natural selection acting on random mutations. But is there any reason to believe so?

Can genetics help? We often hear that chimpanzee and human DNA differ by a mere 1 percent.[4] But humans are radically different from chimpanzees—intellectually, morally, and culturally. Well then, we can be sure that some very important things about us are not found in our DNA. Human exceptionalism is the difference between a clever ape's mind and Einstein's mind.

Is the human capacity for abstract thought located in our neurons? We have roughly three times as many neurons (86 billion) as chimpanzees (22 billion) and gorillas (33 billion).[5] But then elephants, known for animal (and not human) intelligence, have vastly more neurons than we do (257 billion).[6] So the human mind is not merely the outcome of either a vast difference in DNA or the number of neurons.

What happens when we look beyond mammals? Will we see a clear pattern of gradual increases in intelligence until we come to humans? No. Despite their close genetic relationship to humans, apes are not outstandingly smarter than some birds. In 2020, it was admitted at *Scientific American* that young ravens "performed equally well as great apes on understanding numbers, following cues and many more tasks,"[7] though bird brains are organized quite differently from mammal ones.

And then there is the octopus, now celebrated for its remarkable intelligence.[8] Researchers wonder if the octopus developed intelligence independently of the conventional evolutionary explanation.[9] That solitary, short-lived, nine-brained invertebrate[10] is nothing like any vertebrate.

If the human mind depends simply and solely on an evolved physical trait, we have no idea what that trait is.

What About Language?

The difference between the way animals and humans use language is instructive. Many animals communicate extensively. Some animals can understand human communications—but only up to a point. Cats, for example, can distinguish their names and the names of other companion cats among the stream of sounds coming from humans.[11] The cat picks up a *signal*: Something will happen after he hears the sound "Sheldon!" He can't know, however, that he was named "Sheldon" after a sitcom favorite. There is a massive barrier there.

What about the clever birds that learn to mimic human voices? New Mexico State University parrot expert Tim Wright notes that parrots grasp context but not complex meanings:

> When a parrot says "Hello; how are you?" when its owner enters the room, for example, it's "not necessarily interested in your well being" but is mimicking what it hears the owner saying when he or she comes in. In fact, a parrot's understanding of "how are you," is probably "Oh look, someone has come into the room."[12]

That's roughly what we might expect. No animal has the human capacity for the limitless expression of ideas using syntax, grammar, and boundless new vocabulary.

We see this human exceptionalism with numbers too. Researchers have taught fish to count.[13] But that, in itself, is not too surprising; math is fundamental to our universe. Thus, without thinking in any way at all, even *single neurons* in fruit flies can calculate signals for navigating around a fruit bowl.[14] On a larger scale, bees can tell odd from even numbers,[15] and some frogs can keep track at least as far as 6.[16] These abilities don't depend on intelligence; they are likely hardwired[17] rather than learned. All nonhuman capacity for "counting," rudimentary as it is, is perceptual, not abstract. My dog knows the difference between two dog treats and one dog treat because two treats look and taste better than one. But my dog doesn't know, and will never know, that 2 is the square root of 4 or that 1 is the number of multiplicative identity.

What Is Missing from the Animal Mind?

What is missing from the animal mind is abstract ideas. Thus there is no evolutionary explanation of how humans evolved from nonhuman ancestors to apprehend complex numbers,[18] work with twenty-four dimensions,[19] or envision many varieties of infinity.[20]

Some scientists who deny human exceptionalism have tried to teach chimpanzees and gorillas human language.[21] They had limited success. Yes, some apes have learned hundreds of signals. But then a border collie learned over a thousand such signals.[22] Signals are gestures or sounds that point to an object—pointing to food or barking at a stranger, for example, are signals, which animals use in abundance. But signals are *not* language. Unlike signaling, language is abstract, entails syntax and grammar, and enables unlimited flexibility to convey meaning. Only humans do this.

In light of the utter failure of scientists to find genuine language use in nonhuman animals, what do some researchers say now? That "underlying bias and poor experimental designs" account for apes' poor social cognition relative to humans,[23] or that "rather than force apes to learn our language, we should be learning theirs."[24]

Where does this remarkable and uniquely human capacity for abstract thought and language come from? Here's another mystery: When we examine the tree of life, many animals go a long way toward intelligence—but they always stop short of abstract thought. They never reach abstract reason or the language it gives rise to. Whatever is missing isn't in the genes, brain, or body structure. We'd surely have found it by now if it were.

Feelings Versus Questions About Life

Evolutionary psychologist David P. Barash argues that even worms feel pain.[25] But in what *sense* do they feel it? Just as an intellect is more than a survival program, suffering is more than an alarm system.

Human suffering entails more than sensations. It entails profound contemplation—for example, "What did I do to deserve this? Why would God, if there is a God, put me through this?"—that is unique to human beings. No nonhuman animal contemplates such things.

A worm's pain alarm could be sounding in an empty building. As Natalie Mesa has pointed out at *The Scientist*, "Locusts continue to chew leaves as they're being consumed by predators,"[26] which certainly suggests that they do not experience suffering as such. Animals like dogs, cats, and horses clearly do have an emotional *self* that suffers pain and fear and experiences comfort and joy, as we do. But only humans ask, "Why?"

What Mental Qualities Appear Unique to Humans?

The true reason why other life-forms cannot really learn human language is not that we haven't tried hard enough to teach them. As we have seen, what's missing from animals' mental capacities is abstract thinking, an intellectual power unique to the human soul. Hunger teaches an animal to distinguish between one bit of food and two bits of food, but not to wonder about whether a circle can be squared.

Distinguishing between 2 and 4 can be a matter of detecting

signals (less versus more), but distinguishing between 10,052 and 10,054 requires us to think abstractly. Animals can form pictures of things in their minds, but they can't understand the abstract meaning of the pictures. For example, an animal can see a book just as easily as you or I can. But an animal can't read the book and understand what it means. Abstractions—immaterial ideas—are essential for intellectual pursuits. But that sort of thing does not get a fly its fruit or a frog its fly.[27] How then would it evolve?

Animals certainly grieve when those they love die, but they don't grasp the abstract implications—the philosophical meaning—of death. The century-old story of faithful Japanese dog Hachikō beautifully illustrates the point.[28] Akita dog Hachikō had been in the habit of waiting at the Shibuya train station in Tokyo for his human friend, Professor Ueno, to return from the university. One workday in 1925, however, Dr. Ueno died of a cerebral hemorrhage and never returned from the university. Famously, every day thereafter for as long as he physically could, Hachikō went down to the station to wait for him.

Faithful Hachikō, justly honored by a bronze statue today, could not possibly know that his beloved human friend could *never* return. "Never" is an abstraction. It is so powerful an abstraction that in English, for example, we swiftly switch tenses when referring to a deceased person, from "is" to "was." That person's life, however significant, is now in the *past*. We can share with Hachikō what he did understand, faithfulness. To understand what he couldn't grasp, the nature of mortality, is a signature of our humanity.

If demonstrating that there really is no human–animal divide is the goal, legions of researchers toiling over many decades have utterly failed. In fact, their ability to ruminate on the question is itself evidence of the unbridgeable gulf between the human and animal mind: No animal contemplates the gulf, but humans do.

There Is No Such Thing as a Fossil Mind

Nearly two centuries ago, the battle over the human mind split two great thinkers. Charles Darwin (1809–1882) and Alfred Russel Wallace (1823–1913) technically share credit for the theory of evolution by natural selection, but Darwin became the one everyone recognizes. A key reason the two men parted ways—and Wallace was abruptly sidelined—is that Wallace did not agree with Darwin that the human mind had evolved naturally from animal minds.[29]

Many people today would be surprised to learn that there is no good evidence for Darwin's position on the human mind. There is no evidence at all of a progression from inferior human minds leading up to the ones we experience today. Quite the reverse: We find ever more evidence of early human mental abilities. And yet, when we look at the science literature, we find a vast variety of far-flung claims about how humans slowly, perhaps accidentally, acquired a mind. For example, chimpanzees throwing excrement[30] and baby apes' arm-waving[31] are among the many proposed explanations. But shouldn't we wonder why, if excrement-throwing and arm-waving are the secret to human exceptionality, these practices did nothing similar for chimpanzees, who, millions of years later, still lack the capacity for abstract thought that comes so naturally to humans?

Neanderthals long filled the role of not-quite-us, to the point where "Neanderthal!" became a term of abuse. But then we dug up much evidence of their use of pigments, decorative shells,[32] feathers as decorations,[33] carvings,[34] tools,[35] and artwork.[36] In other words, Neanderthals never succeeded in being as stupid as human evolution theory requires. Artistic creation is a hallmark of humanity's capacity to think abstractly. And none of the other ancient humans excavated seem to be the not-quite-human either. George Washington University paleobiologist Bernard Wood has admitted, "The origin of our own genus remains frustratingly unclear. Although many of my colleagues are agreed regarding the 'what' with respect to Homo, there is no consensus as to the 'how' and 'when' questions."[37]

Well then, if man is biologically an animal but cognitively so unlike an animal, the obvious implication is that some aspect of the human mind is not biological. One element of our known history that illustrates this fact is the difference that big abstractions—things like death and religion—made in early human history.

Consider the lovingly prepared grave of a child, called by archaeologists "Mtoto," from seventy-eight thousand years ago in what is now East Africa:

> Special care was taken in burying Mtoto, Martinón-Torres said. For instance, the skull base and attached neck bones had collapsed away from other back bones in a way suggesting that the child's head originally lay on a support or pillow that had decayed. Rotation of a collarbone and two ribs suggested that Mtoto's upper body had originally been wrapped in a shroud.[38]

Then there are the ninety-thousand-year-old graves at the Qafzeh Cave[39] in Israel:

> Modern behaviors indicated at the cave include the purposeful burials; the use of ochre for body painting; the presence of marine shells, used as ornamentation and, most interestingly, the survival and eventual ritual interment of a severely brain-damaged child.[40]

The researchers are sure that the twelve- or thirteen-year-old in the Qafzeh Cave had been looked after for a long time, because 3D imaging showed serious skull fractures that had been sustained much earlier than the tooth development at death. The researchers also noted that "the child appears to have received special social attention after death, as the body positioning seems intentional with two deer antlers lying on the upper part of the adolescent's chest, likely suggesting a deliberate ceremonial burial."[41]

Erella Hovers of the Hebrew University of Jerusalem, who directed the Qafzeh project, comments that "many symbolic behaviors that are considered modern existed for a time [before the Upper Paleolithic] and then disappeared, to be reinvented time and again."[42] Abstract thinking can indeed resurrect ideas; we've seen that happen in modern history too.

Most remarkable of all, perhaps, are the roughly six hundred limestone balls (spheroids), the size of plums, found alongside stone axes at a site in northern Israel, dating from 1.4 million years ago. Archaeologist Antoine Muller of the Hebrew University of Jerusalem Institute of Archaeology comments, "It appears that hominins 1.4 million years ago had the ability to conceptualize a sphere in their minds and shape their stones to match."[43] If they had the appreciation of symmetry and beauty with which the study authors credit them, all we can say is, technology has greatly improved since then. And yet the human mind that sought the perfect sphere remains the same.

And religion? Contemplation of spiritual matters, of our purposes in life and of our ultimate destiny, is a hallmark of abstract thought. The most remarkable known religion before the dawn of writing is surely the recently discovered worship site in Anatolia, Göbekli Tepe, from nearly twelve thousand years ago. As Charles C. Mann wrote in *National Geographic*, it is

> made not from roughly hewn blocks but from cleanly carved limestone pillars splashed with bas-reliefs of

animals—a cavalcade of gazelles, snakes, foxes, scorpions, and ferocious wild boars. The assemblage was built some 11,600 years ago, seven millennia before the Great Pyramid of Giza. It contains the oldest known temple. Indeed, Göbekli Tepe is the oldest known example of monumental architecture—the first structure human beings put together that was bigger and more complicated than a hut.[44]

We can have no idea, of course, what the religion was—not its cosmology, nor its teachings, nor its hierarchy. What we can know is, in the words of a researcher in 2011, "Twenty years ago everyone believed civilization was driven by ecological forces. I think what we are learning is that civilization is a product of the human mind."[45]

He's right, but the capacity for the abstract reasoning that underlies these developments is inherently immaterial.[46] Can a thing that is by its nature immaterial even *be* a product of evolution?

The Human Mind Did Not Evolve

How can Darwin's theory of evolution be applied to the mind? The theory posits that changes in DNA that favor survival and reproduction will cause some life-forms to outcompete others. Over time, surviving life-forms would then evolve into very different ones, cows into whales, perhaps. But there is no DNA for abstract human thought, for reason and free will. The human mind's intellectual powers are not material, as the neuroscience we have discussed in this book shows. They may or may not even aid survival in Darwin's sense; misused— think of nuclear weapons and chemical and biological warfare—they often contribute to destruction. Most of the world's great literature addresses the dangers posed by human schemes and purposes. While human technology has evolved a great deal, there is no evidence that the minds behind it have evolved—or even that they can.

Not only do we have no adequate theories for the mind's natural origin from earlier animal minds, we don't even have the *basis* for such a theory. Darwin's theory of human evolution is a ladder with no lower rungs. The evidence available shows that the specifically human mind, with its capacity for abstract thought and free will (that is, the human soul), could not and did not evolve.

CHAPTER 11

What Does It All Mean? Neuroscience Meets Philosophy

I BEGAN MY MEDICAL career forty years ago, fascinated with neurosurgery and in love with science. I was convinced that a deeper scientific understanding of the brain would reveal the secrets of the mind. Now, seven thousand brain operations later, I am still fascinated by my work and I still love science.

But both science and experience have taught me something quite different from what I had expected. I realize now that while the textbooks are written by leading neuroscientists, these basic scientists lacked experience with real people. Scientists rarely enter the operating room. Most neuroscientists have never seen a living human brain and have never examined a patient following brain surgery. Much of their neuroscience is detached from real life, done in a laboratory or under a microscope, not in an operating room or a clinic. My textbooks didn't always give me accurate information about the brain and the mind.

As I shared earlier, I have talked to people while removing tumors or large damaged parts of their brains. I have examined many patients in my clinic while looking at brain scans that showed large parts of their brains missing or mangled—yet the patients were perfectly lucid. I've had thousands of patients with damaged or disfigured brains yet pretty good minds. Thus, I learned that there are parts of the brain that don't seem to matter in everyday life—parts that can be removed or injured without harm.

When I studied the work of fellow neurosurgeons and neuroscientists, I learned that the ability of the mind to reason never shows up in brain seizures and can never be evoked by stimulating the brain with an electrical probe. Remarkably, while a shot of whiskey or a blow to the head can impair reasoning, reason itself doesn't seem to come from the brain—at least not in the same way that movement, sensation, emotion, and memory come from the brain. I learned that surgically splitting the brain hemispheres apart causes something remarkable—the ability to perceive things is subtly altered. Astonishingly, people who have had split-brain surgery can even compare things they see when no part of the brain sees all the things they are comparing! There is a part of the mind—the ability to bring our experiences together in ways that make rational sense to us—that cannot be split by a surgeon's scalpel.

I learned that the ability to choose freely doesn't seem to come from the brain itself, although unconscious temptations that emerge from the brain bombard us continuously. I also learned that some people can think clearly even in the deepest level of coma and, using modern technology, even answer questions. I learned that near-death experiences are surprisingly common, and that some of these experiences happen when doctors have confirmed that the brain was not working at all. In the case of Pam Reynolds, her brain could not possibly have been working at all when she had her near-death experience—her heart was stopped and her brain was drained of blood!

The truth is that the brain and the mind are related in unexpected and even odd ways that are not told in the textbooks. How can we understand this? We've already seen that materialism and its theories are inadequate to explain the mind. So what *can* explain it?

I am a committed scientist, but I don't think the best way of understanding what science has revealed about the human mind is through science alone. Science cannot interpret itself. We must also draw on philosophy and theology. I've been sketching out parts of my approach: Here is a fuller explanation.

What the Ancient Philosophers Knew

In my own quest for understanding, I discovered that ancient philosophers and theologians such as Aristotle and Thomas Aquinas had pondered many of the same questions I had, although they lived long before modern neuroscience and neurosurgery. Yet they understood that we have souls, and their descriptions of the powers of our souls are remarkably like what modern neuroscience is revealing to us in the twenty-first century.

Aristotle began trying to understand human nature by asking the most basic question about nature itself: What things exist, and how can we know them? He used a term for the things that exist—rocks, trees, animals, and man—*substances*. In his view, we can know two things about a substance: its *form* (intelligible aspects such as size, weight, color, internal workings, and so on) and its *matter* (its individuation, that is, what makes it *this* thing rather than the identical thing next to it). For example, the form of a tree is what we can see about it *as a tree*, in general. Without knowing many more details, we would know it was a tree. The matter is what distinguishes the particular tree in the front yard from a nearly identical one in the backyard.

The form of any living thing is its soul and the matter is its body. The soul of a live tree, dog, or man is all the things which make that being alive instead of dead. It is the difference between a living organism and a corpse.

Recall that Aristotle identified three types of soul: Plants have *vegetative* souls, which enable them to grow, feed, and reproduce. Animals have *sensitive* souls. The sensitive soul has all the powers of the vegetative soul but in addition enables animals to feel things, move, form mental images, remember those images, and experience emotions. Human beings do all these things as well, but, of course, we also have *rational* souls, which give us powers animals and plants lack—the powers of reason (abstract thought) and free will. Aristotle called these powers intellect and will.

There Is No Organ for Reason or Willpower

Vegetative and sensitive powers are closely linked to the matter of the body—we take nourishment into our stomachs, see with our eyes, and move with our muscles. But our rational powers of intellect and will have no organ—they are not generated by any part of the body, not even the brain. The normal working of our bodily powers—vision, hearing, emotional states, memory, and so on—is necessary for the normal function of the intellect and will, *but not sufficient*. That is because intellect and will are immaterial powers of the human soul.

To visualize this, consider a triangle in the abstract again—but this time, let's go a little deeper into the concept. It is a closed-plane figure with three straight sides and interior angles that add up to 180 degrees.

But how can you *know* this? After all, you have never *seen with your own eyes* a plane figure—that is, a figure on a completely flat surface (all physical surfaces have at least subtle irregularities). You have never seen a perfectly straight line. If you have measured the interior angles of a physical triangle with a protractor, they never add to exactly 180 degrees. This inescapable imperfection in the world is evident in everything that can be expressed by mathematical concepts.

Although we can easily conceive of perfect circles or absolutely straight lines, perfect conceptions cannot arise from our very imperfect perceptions. There is an *unbridgeable gap* between material powers of the human mind (movement, sensation, memory, and emotions) and immaterial powers of the human mind (intellect and will).

Some might object, of course, that real-world triangles are close enough to perfection that our minds can just make the conceptual leap. Yes, but to make that leap, we must start by assuming that the idea of perfection exists in principle. Countless abstract concepts are like that; they *can't* be physical. Logic, mercy, justice, perfect circles, perfectly straight lines, and a host of other concepts can have no physical existence yet are quite real in our minds. There is no getting away from the immaterial aspects of the human soul.

What It Means to Say That We Have a Human Spirit

Thomas Aquinas, along with some other Christian theologians and philosophers, began to adopt Aristotle's understanding of the human soul. But they made one important change. They kept his psychology largely intact, but they thought in terms of the human *spirit*, in the Christian theological sense, as the origin of the immaterial powers of intellect and will. That is, we have spiritual souls because we are created in God's image and thus share in His spiritual nature in some senses. The immaterial human spirit—the intellect and will—acts through the body to accomplish its purposes.

Adopting Aristotle's way of thinking about the soul enabled Christian theologians to express theological concepts in the terms used in a widely accepted philosophy. But the move was highly controversial at first.[1] Aristotle was, after all, an ancient Greek philosopher, thus a *pagan*!

Aquinas had a strategy to get around this problem, though. As philosopher Edward Feser, an Aquinas specialist, points out, Aquinas "held that the existence of God, the immortality of the soul, and the content and binding force of the natural moral law could be established through purely philosophical arguments (as opposed to an appeal to divine revelation)."[2]

In short, Aquinas sought to show that, whether one accepts the authority of Scripture or not, denial of the existence of the soul is a quick route to illogical beliefs. For example, as we have seen, materialist philosophers often attempt to explain away consciousness,[3] *using* consciousness!

This approach to philosophy was generally consistent with Christianity. As Genesis 2 puts it, man is a hybrid, a composite of earth and spirit—God breathed into man, made of dust, the breath (the spirit) of life. Our Creator is a Spirit, and it is in this sense that we are created in His image, as both spiritual creatures like Him and material creatures like His creation. Thus our lives are rooted in biology, yet we can contemplate our biology, among other things, abstractly. No other animal does this.

From the perspective of traditional Christian philosophy, we humans are the only known composite creatures—atoms, rocks, plants, and nonhuman animals are all matter, and angels are all spirit. Of course, just as the immaterial human spirit can cause material effects, it can also be acted on by material causes, as a few ounces of alcohol or a head injury attest. But in its essential nature, it has powers that transcend matter. Because we bridge the two great domains of existence—spirit and matter, heaven and earth—our unique place in creation is worth a closer look. It will give us a clearer picture of the nature of the soul.

What Does It All Mean? Neuroscience Meets Philosophy

Finding and Understanding Our Souls

Some ask, "*Where* is the human soul?" Is it in the brain, the heart, or in an obscure little organ like the pineal gland (as the thinker Descartes believed)?[4] That's a bit like the neuroscientists today who are looking for the "consciousness spot" (or circuit) in the brain.

Our soul is a spirit, and our spirit is a soul. The spiritual human soul is the sum of all of the powers that make us living human beings. The spiritual soul is the beating of our heart, the metabolism of our cells, the exchange of oxygen and carbon dioxide in our lungs, our ability to move, to perceive, to remember, to have emotions, to reason, and to have free will, along with innumerable other things that living human beings can do.

It is wrong to think of our spiritual souls as filling our bodies like translucent ghosts of an equivalent shape and size, like Patrick Swayze in the movie *Ghost*. Our spirit, by its nature, has no physical location per se—it is not matter and thus does not occupy space. Rather, a spirit can be said to be where it *acts*; our spirit is in our body in the sense that it acts there, carrying out the processes of life.

The Difference That the Human Soul Makes

To see how the spiritual soul in humans differs from the wholly material soul in animals, consider what would happen if a dog and a man saw an enslaved woman working in a field. The dog sees the woman, her clothing, and the tools she is using, and feels the same hot sun as she does. If he remembers her later, he may feel emotions ranging from fondness and friendliness through to dislike or fear. The man observing the same woman has roughly the same range of material mental powers of perception as the dog. But unlike the dog, the man can also *contemplate* slavery, consider what it means.

The Immortal Mind

Unlike a dog, a man could think about what it must mean to be torn from your ancestral home to work as the property of another human being and to have your family sold at auction. The man *understands* what he is experiencing and arrives at a conclusion about the abstract meaning of it all. He might then read (or even write) a book about slavery (Frederick Douglass),[5] fight against slavery (John Brown), or lead a nation to end slavery (Abraham Lincoln). Conversely, the same man may choose to defend slavery, purchase a slave of his own, or fight in battle to retain the "peculiar institution," as Lincoln called it.

Of course, that same man may instead choose to act only on his emotions, as a dog would, without reflecting on what any of it means. But that too is a choice. The dog cannot contemplate slavery and thus cannot choose either of these moral opinions or even refuse to form one. These immaterial powers of the human soul are part of what it means to say that man is a *spiritual animal*.

Human knowledge is qualitatively different from animal knowledge. My dog can remember the smell of bacon more passionately and much more acutely than I ever will. But he will never wonder whether too much bacon is bad for him, let alone whether producing bacon causes too much suffering to other animals.

Because our human souls can have this abstract type of knowledge, we can think of God without seeing, hearing, or touching Him. Just as we can contemplate God's laws of physics, we can contemplate His laws of morality. God put His physical laws into our bodies—laws of chemistry and physics and physiology—and His moral laws into our souls. There is a difference between the two: We can't choose to violate His physical laws like gravity, but as spiritual beings with free will we can choose to violate His moral laws. But of course it is still His universe, and violation of His moral laws has consequences, which are not as immediate, but are no less certain, than gravity.

What Does It All Mean? Neuroscience Meets Philosophy

The Sense Organs of the Human Soul

One way of thinking about all this is that the spiritual human soul has "sense organs" just as the body does. Just as the eye is directed to light, the intellect is directed to the truth, and the will is directed to the good. It is the free will—directed to what the intellect understands as good—by which we choose. Conscience is the eye of the spirit.

This ancient understanding of the spiritual human soul is today supported by a century of neuroscience.

Neuroscience's Window on the Soul

Consider the rational (that is, spiritual) aspects of the human soul that we have learned about from neurosurgical experience and from modern neuroscience. The spiritual soul enables us to know, love, and act—this is obvious from both neuroscience and from everyday life. The spiritual soul has no parts—it cannot be split by a surgeon's knife during split-brain surgery. While it seems at first that the spiritual soul exists in space and time, on deeper thought that seems not to be the case. We might say that the spiritual soul is located where and when it acts—in the body, for example. But when we consider it carefully, whereas the body obviously has a location and time, the spiritual soul itself has no real location or time—it makes no sense to ascribe location or time to a spiritual soul that emerges into another kind of life after death. People with near-death experiences often have life reviews in which they experience their full life vividly as if it were happening all at once. And last, and perhaps most importantly, the spiritual soul abides. By that I mean that it survives massive brain damage and even death. It doesn't "snuff out" like a candle flame, even when that brain and body are damaged or dead.

Our rational soul has the same basic properties, on a vastly limited scale, as do angels and even God. The difference between our spirit and that of angels and of our Creator is one of scale and of kind—our power of reason is cumbersome and sequential, whereas angelic and divine reason is incisive and instantaneous and, for God, infinite. Spirits know and love and act, spirits have no parts and cannot be split, spirits have no location and are not trapped in time. And spirits abide—they cannot disintegrate and die.

Finally, we are not souls merely trapped in bodies, destined to be freed from physicality. A complete human person is soul and body inseparable.[6] Paul stresses in 1 Corinthians 15 that when souls rise to life immortal, they are given a new body, suited to their state of being. A human is created to be a hybrid being—a spiritual soul animating a material body—and ultimately a perfect one.

This is at the heart of the mystery I discovered in surgery and in learning and contemplation about the mind and brain—we are hybrids, composites of spirit and matter, with souls that bridge the spiritual world and the material world. We are mineral (in our bones), vegetable (in our growth and nourishment), animal (in our movement and sensations), and spiritual (in our reason and free will). We humans embody all elements of creation.

CHAPTER 12

And This All Men Call God

AT THE BEGINNING OF this book, I shared with you how I started my journey to understand the relationship between science and the soul. I revealed that it was part of my larger journey to God. Here I want to connect the dots between my two quests.

I knew enough philosophy to know about the principle of *sufficient reason*, a maxim of classical philosophers that everything that exists in the natural world necessarily has a sufficient reason for its existence, no matter how difficult it may be to discover that reason. Nothing exists or happens for no reason at all. I understood that our spiritual souls have a reason for their existence.

Where did our spiritual souls come from? As I contemplated this deepest of questions, I kept coming back to my hauntings and my thoughts about the mansion. How did I come to be in this world? What—or Who—was the cause of me and the cause of the universe? The Christian faith I knew from childhood (but didn't really practice or believe yet) taught that creation came from God, but this seemed just dogmatic, and the truth of it didn't yet resonate with me.

Maybe you have struggled with these questions too. It turns out we are not the first humans to do so.

Göbekli Tepe, the immense temple complex discovered in Anatolia in 1994, is the oldest known large religious structure, as mentioned in chapter 10. Its immense pillars and beautiful stone carvings, dating from 11,500 years ago, were exhausting, multigenerational work. And we'll probably never know why, after about 1,500 years, it was abruptly abandoned. Loss of faith? All we know is that humans, so far as we have any record, have always thought deeply about life and death—and, inevitably, about who or what started it all.

For convenience, we will call it belief in God, even though some religions (Buddhism, for example) offer a different account of this conviction. Can we prove anything about such a Mind behind the universe? Or is it all just a matter of faith? Surprisingly, we *can* prove some things.

Unfortunately, science has not always been much of a help here. Many researchers seem committed to explaining such insights away. Thus, science media are rife with trivial, unsatisfactory accounts of the origin of a perennial, worldwide belief in a divine origin of our world—and a divine context for our lives.

Does Belief in God Help Humans Survive or Is It Just a Glitch?

For over 150 years, influential social scientists have argued that truth or falsehood is irrelevant. Darwinism, they say, is the answer. Belief in God makes humans more likely to survive and pass on their genes. Thus a key question is framed: Does religion increase our evolutionary fitness?[1] Or is it just a glitch that doesn't kill us off?[2]

Nineteenth-century eugenics advocate Francis Galton (1822–1911) thought that religion helped fitness: "Between two barbaric nations, the one that was the more superstitious of the two would generally be the more united, and therefore the more powerful."[3] In our own day, evolutionary psychologist E. O. Wilson (1929–2021) argued in a similar vein: "The individual is prepared by the sacred rituals for supreme effort and self-sacrifice. Overwhelmed by shibboleths, special costumes, and the sacred dancing and music so accurately keyed to his emotive centers, he has a 'religious experience.'"[4]

But others have put it down to a glitch in the human way of thinking: Yale University psychologist Paul Bloom offers a "glitch" view, based on his study of small children: "One: human beings come into the world with a predisposition to believe in supernatural phenomena. And two: this predisposition is an incidental by-product of cognitive functioning gone awry."[5]

Bloom, an atheist, can make such pronouncements without anyone asking how Darwinian evolution predisposes *him* to think. Atheists seem to offer few theories about how atheism evolved—the purpose of atheist theory seems to be to explain *away* the beliefs of others. Their bedrock assumption is that there is not and cannot be evidence for the existence of God.

Does the Atheist Approach to Religion Advance Science?

One result of these Darwinian fables has been a huge investment in research that borders on the ridiculous. For example, a massive and widely celebrated study, published in *Nature* in 2019, argued that the mere existence of large societies spurs belief in "fire-and-brimstone gods." It was retracted in 2021 due to recognized problems with the data.[6] But most such claims go unchallenged, and many are widely disseminated in popular science media.

The fitness thesis and the glitch thesis can't both be right, but they can certainly both be wrong. If religion arises from real awareness of a divine presence in the universe and in our lives, they *are* both wrong. And yet no matter how well God's real existence fits the evidence, in the view of many scholars today, accepting it just couldn't be *science*. They assume without evidence that atheism somehow provides a clearer perspective.

At the same time, every day many people change their lives dramatically because, they say, they encountered God, a real Person—the only absolutely real Person. When such an experience results in a life changed for the better, it provides powerful testimony for others. But while a changed life is significant evidence, it is, after all, usually a private event. How does it fit into an overall framework of reality? That's where we look to the public lines of evidence from science and philosophy.

A Logical Reason for Thinking That There Is a Mind Behind the Universe

Almost all cultures worldwide have a creation story. As reasoning beings, we find it natural to ask how things got started, why the world is the way it is, and how or whether it will change or end. We reasonably assume that whatever created the universe cannot logically be just another part of it. It makes no sense to say that the universe caused itself.

We make the reasonable assumption that the human soul—the principle that makes each of us a living person with the unique power of reason and free will—could only be created by a Person with a Mind capable of such powers. Thus, the author of Hebrews writes, "By faith we understand that the universe was formed at God's command, so that what is seen was not made out of what was visible." (Hebrews 11:3).

It is not an assumption specific to one culture: In Islam, "The act

And This All Men Call God

of creation, therefore, is frequently described as a way of drawing the reader into thinking about the order of all things and the All-Knowing Creator Who is behind it all."[7] In Hinduism, "the universe was created by Brahma, the creator who made the universe out of himself."[8] Buddhists don't typically think in terms of divine creation, but they do have a creation story of a sort: Our present universe is the latest in an infinite past series.[9] Thus infinity is itself the creator. Generally, however, the most widely followed traditions assume the existence of a divine Mind outside the created order. Is this a correct intuition supported by evidence and reason?

Science Evidence for a Mind Behind the Universe

Science is replete with powerful evidence for the existence of a Creator. Consider the fine-tuning of the universe for life.[10] Most physicists, irrespective of what else they believe about God, simply accept the fine-tuning as a fact. Physicist and broadcaster Paul Davies, who takes a generally agnostic position on God, told a BBC audience, "The really amazing thing is not that life on Earth is balanced on a knife-edge, but that the entire universe is balanced on a knife-edge. You see, even if you dismiss mankind as just a mere hiccup in the great scheme of things, the fact remains that the entire universe seems unreasonably suited to the existence of life—almost contrived—you might say a 'put-up job.'"[11]

Fermilab senior scientist Don Lincoln, who dismisses any idea of a divine Mind ("It's not all that different from saying 'just because'"), nonetheless offers a number of remarkable facts from his own discipline, concluding, "Scientists know dozens of examples of little changes that would radically alter the Universe. A small change in the laws of gravity could have caused the Universe to collapse into a black hole immediately after it came into existence. Alternatively, gravity could have been too weak for stars and galaxies to form. Either way, we wouldn't exist."[12]

And, of course, religious scientists assimilate the facts into their own traditions. Cosmologist George Ellis, a Quaker, told a recent cosmology conference, "Amazing fine tuning occurs in the laws that make this [complexity] possible. Realization of the complexity of what is accomplished makes it very difficult not to use the word 'miraculous' without taking a stand as to the ontological status of the word."[13] Nobel Prize–winning physicist (1978) Arno Penzias, who helped establish the Big Bang theory, offers a thought from the Jewish tradition: "Astronomy leads us to a unique event, a universe which was created out of nothing, one with the very delicate balance needed to provide exactly the conditions required to permit life, and one which has an underlying (one might say 'supernatural') plan."[14] He added, "My argument is that the best data we have are exactly what I would have predicted, had I had nothing to go on but the five books of Moses, the Psalms, the Bible as a whole."[15]

But Can It All Be Explained Away?

Iconic physicist Stephen Hawking (1942–2018), an atheist, offered an alternative explanation: "In the same way that the environmental coincidences of our solar system were rendered unremarkable by the realization that billions of such systems exist, the fine-tunings in the laws of nature can be explained by the existence of multiple universes."[16] That is, there are countless other universes out there and ours just happens to work this way. The trouble with his explanation is that we have no evidence for the existence of even a single other universe, let alone multiple ones. In any event, what if there were multiple universes and they were all fine-tuned? Such an account of fine-tuning seems to come either from nowhere or from a reluctance to accept the origin of our own universe in the activity of a divine Mind. Fine-tuning of *any* sort, speculations about multiple universes notwithstanding, inherently points to a Mind of vast intelligence and power.

One thinker who grasped the weakness of popular claims that the cosmos could somehow evolve without a divine Mind was C. S. Lewis (1898–1963): "We are taught from childhood to notice how the perfect oak grows from the acorn and to forget that the acorn itself was dropped by a perfect oak. We are reminded constantly that the adult human being was an embryo, never that the life of the embryo came from two adult human beings. We love to notice that the express engine of today is the descendant of the 'rocket'; we do not equally remember that the 'Rocket' springs not from some even more rudimentary engine, but from something much more perfect and complicated than itself—namely, a man of genius." Lewis concluded, "On these grounds and others like them one is driven to think that whatever else may be true, the popular scientific cosmology at any rate is certainly not."[17]

Arguments from Reason for the Existence of God

When we showed that the immortal human soul exists, we used two types of argument, arguments from evidence and arguments from reason. First, there is neuroscience evidence for the existence of the soul. Second, reason shows us that the soul is the type of entity that is immortal by its very nature. When we discuss the existence of God, we can use the same approach. Fine-tuning, like neuroscience findings, is evidence that is available to everyone. Of course, many of us believe that God exists for a more personal but still evidence-based reason—an experience of His goodness and power. But that evidence is not obvious to everyone.

But there are also sound arguments for the existence of God based on reason. Earlier in my life, I turned to pagan philosophy, particularly that of Aristotle. Aristotle taught that, to understand how anything in the world is what it is, we need to understand its four causes. A common illustration in modern times is the statue of David by Michelangelo. To understand all there is to know about the statue, we need to know its material cause, which is the marble from which it is made. We need to know its formal cause, which is the shape. We need to know its efficient cause, which is Michelangelo and the tools by which he made the statue. And we need to know its final cause, which is the purpose for which it was made—for example, to provide a decoration of great beauty (and, because Michelangelo had bills like we do, to get paid!).

Aristotle proposed that one cause—the final cause—was more important than the other causes, because it is the *purpose* for the existence of the statue. The purpose of the statue dictated that it be made of beautiful Carrara marble (material cause), that it depicts David (its formal cause), and that it shows the passion and hard work of Michelangelo (its efficient cause). The final cause is, in Aristotle's terms, the Cause of Causes. Final cause, which is purpose, pulls all the other causes along with it.

Aristotle and philosophers and theologians who followed in his footsteps called final causation *teleology*, from the root *telos*, which means "goal" or "purpose." Teleology is the study of purposes in the world. And that leads naturally to the logical arguments for the existence of God.

The Five Ways of Knowing That God Exists

A number of logical arguments for the existence of God were developed by Aquinas. As a priest and a Dominican friar, he believed, of course, that some truths about God can be known only by revelation. But others, he sensed, can be inferred by careful reasoning from first principles.[18] When discussing the existence of God, he preferred to argue from the principles of reason. Thus he offered five ways we can know that God exists, sometimes ending with "this all men call God." That is, if these statements about the nature of reality are reasonable, cultures across the world are correct in recognizing God's existence as a logical conclusion.

The First Way is the Prime Mover argument. It was originally proposed by Aristotle, but Aquinas adapted it to Christian theology. It's quite simple. For movement, there must be a mover. By "movement," Aristotle meant change of any sort, not just change of location. It could be a series of simultaneous changes. The classic example is of a man using a long stick to move a leaf on the ground. The final change is the movement of the leaf, and the series of simultaneous changes is the movement of the various parts of the stick. In order to account for the changes—for the movement of the stick and the leaf—we must suppose a hand of some sort on the end of the stick. The stick won't move itself, no matter how long it is. So we can't propose an infinitely long stick as an explanation for movement. In this case, we know that a man's hand moves the stick, the man's muscles move the hand, the man's nerves move the muscles, the man's brain moves the nerves. The energy taken by the food the man eats moves his brain, the energy in the food comes from sunlight, and so on, all the way back to God.

God changes things but is not Himself changed. It is not in His nature to change. Thus, requiring nothing to initiate a change in Himself, He can be the ultimate origin of a series of changes. One way of putting this is to say that God is a *necessary being*, in fact, He is the only necessary being. He must exist if anything else is to exist.[19]

The Prime Mover argument, in its most rigorous form proposed by Aristotle and Aquinas, is logically very strong and has never been convincingly refuted. In formal language, it states that a series of instrumental changes cannot go to infinite regress. It requires a First Changer (or Prime Mover) who gets the process of change started. That Prime Mover, which must be outside the process, is what all men call

God. Aquinas considered the First Way to be the strongest argument of the five.

The Second Way or First Cause argument also looks at the ultimate origin of things. But it focuses on causes of events rather than on movement/change as such. It follows the same logical sequence as the Prime Mover argument—a chain of instrumental causes and effects cannot go to infinite regress. Therefore there must be a First Cause, an uncaused cause at the origin of all causes in a chain. And that is what all men call God.

In the Third Way, Aquinas looks at existence itself. Existence is understood, as in the first two ways, as a chain of causes and effects. It happens in a causal chain—one thing exists because something else caused it to exist, and that something else itself had to have a cause for its own existence, and so on. Aquinas argued that, like a chain of changes and a chain of causes, the chain of existence must have a Necessary Existence that does not depend on anything else for its own existence. And this is what all men call God.

The first three ways are called the Cosmological Arguments, because they depend on the order of chains of changes, causes, and existence in the universe. They show that each chain needs something outside of itself to get started.

Looking Past the Origin of Things

But now another question arises. So far, the power that started the universe is seen as the Prime Mover, an Uncaused Cause, and a Necessary Existence. But is it anything more? Does that power within which we live and move and have our being (Acts 17:28) have other qualities that reason itself—in addition to revelation or evidence—would suggest?

In the Fourth Way, Aquinas looks at the question of greater and lesser *perfection* of being. Everything in nature features degrees of perfection. For example, there is no perfect circle in nature, but some circular things come much closer to perfection than others. We could say that they participate more fully in circularity.

Perfect circularity is a real concept used every day in mathematics. But it is a different sort of reality than circular objects that have the potential to be more or less perfect without ever achieving actual perfection. Potential *means* not yet actual. But then, to avoid violating the law of noncontradiction, we must assume that there is an actual perfection in which the less perfect circles incompletely participate. We can envision that fully actual perfection in a number of areas—it could be pure Goodness, pure Truth, pure Nobility. And that is what all men call God.[20]

The Fifth Way looks specifically at the order that we find in nature. Nature follows consistent patterns—the sun rises and sets, leaves change color in the fall, dropped stones fall to the ground (and not to the sky), and so on. Aquinas draws attention to something we take for granted but rarely pause to reflect on: Countless objects in nature, including life-forms, behave in very predictable ways so as to organize a vast complex structure, without any awareness of that fact on the part of the objects themselves. This implies that there is a Mind that oversees the patterns in nature.

The Fifth Way is a kind of design argument, but it depends on the regularity of nature and not necessarily on the complexity of nature. The analogy Aquinas gives is that of arrows we might see flying through the air and consistently hitting a target. Even if we couldn't see where the arrows came from, we would correctly infer the existence of an archer. Aimed arrows *presuppose* an archer, and consistency in nature presupposes a designer. It must be, he held, a divine Mind that is not a part of the natural order. Rather, it creates and upholds the natural order and enables things in nature to fulfill natural goals. "By wisdom the LORD laid the earth's foundations..." (Proverbs 3:19). And that is what all men call God.

Why Is Design in Nature So Controversial?

Today, we know much more about the design in nature than was known in Aquinas's time. However, Darwinian evolutionists argue that apparent design does not support a Fifth Way argument. Life-forms, they say, show complex development and behavior simply because they are more likely to survive and breed if they do.

There is a fatal flaw in their argument: The mere fact that a life-form might thrive better if it develops a given ability does not provide that life-form with any power to develop it. The survival of the fittest doesn't explain the *arrival* of the fittest. The fact that well-designed organisms survive does not account for the origin of that design. The astonishing design evident in living things is vastly beyond the meager creative power of Darwinian just-so stories, and provides further evidence for a divine Mind. Consider just three examples:

The origin of life—in all its dazzling complexity—4.5 billion years ago is chemically a seeming impossibility, yet it happened. And no explanation that rules out a Mind behind the universe has ever succeeded. A survey article in a science journal noted in 2020 that "origin of life (OoL) is one of the major unsolved scientific problems of the century. It starts with the lack of a commonly accepted definition of the phenomenon of life itself." The article then notes that the field features "highly heterogeneous" research and "strongly opposed views which potentially hinder progress."[21] The most likely cause of the centuries-old impasse is a proverbial "elephant in the room," in this case, the design of life no one wants to acknowledge.

Then there is the Cambrian explosion, the sudden proliferation of complex life-forms a little over half a million years ago. It is, the journal *Nature* tells us, "the most significant event in Earth evolution,"[22] giving rise to most major life groups. Yet if we inquire about a cause, we are told that it was the sudden appearance of a "range of interacting biotic and abiotic processes,"[23] which is true but hardly very explanatory. That's another elephant lurking in the room.

Finally, there is the origin of the human mind. It has, as we noted earlier, no history. Unlike the animal mind, it is capable of understanding and carrying out both great good—and great evil. Physicist Eric Hedin offers, "I would submit that nowhere in the animal world do we see evil that comes anywhere close to comparing with the unfortunate depths of evil displayed by humanity throughout our recorded history."[24] Of course, animals do what they do in order to eat, mate, or defend themselves. It takes a human mind to have greater goals, including evil ones, and the free choice to carry them out. That is part of the mind that has no history. It is most likely a special creation. No other explanation seems plausible.

The idea of a divine Mind behind the universe and life-forms is a controversial topic today because of the presumption of atheism, the default position of current science, not because of the direction of the evidence from nature.

The Moral Argument

Another very powerful argument for God's existence is the undeniable existence of moral law. We humans are intrinsically aware of moral principles—it is wrong to lie, steal, rape, murder, and so on. Of course, we don't always live up to this moral law, but we always feel it pulling on us almost like a force toward moral behaviors. Even a career criminal objects to being cheated or assaulted—"That's not fair!" This internal sense of the moral law cannot be mere human opinion, because it is imposed in various ways by civilizations across the globe as something more than mere personal preference. It amounts to an implicit recognition that the moral law is a part of the toolkit of every human being.

Immanuel Kant (1724–1804) expressed this consequence of God's sovereignty beautifully in his *Critique of Pure Reason* (1781) when he wrote, "Two things fill the mind with ever new and increasing admiration and awe, the more often and steadily we reflect upon them: the starry heavens above me and the moral law within me."[25] The moral law is, in this sense, written in human hearts and is powerful evidence within us of God's existence and sovereignty.

Materialist Atheism's Achievements

What does the presumption of atheism, the default position of current science, offer to replace a divine Mind? We are offered a bewildering variety of proposals about the universe, including no boundaries and imaginary time.[26] We are also told that life must have arisen by "self-assembly."[27] Yet self-assembly is precisely what we never see in the world around us. As Pasteur showed, all life comes from life.[28]

And consciousness? It is, according to Ralph Lewis, offering a conventional view, a "spontaneous, bottom-up, self-organising phenomenon of complexity, with no external cause required."[29] Note the "self-organising" claim. Why do we never see consciousness merely "self-organising" in the vast sea of life around us? Conscious beings arise only from other conscious beings. It is astonishing that materialists would expect us to reject a divine Mind behind the universe and life-forms when they have no reasonable, evidence-based proposition to offer as an alternative.

A stark truth emerges. The presumption that atheism is the official position of the sciences today is not evidence-based; it is a mere *presumption*. It is well expressed by renowned Oxford chemist Peter Atkins: "A scientific atheist holds that the domain of science is the physical world, but considers there is no other variety of world, and that the 'spiritual' is an illusion generated by a physical brain."[30] His approach is becoming less and less plausible in light of the evidence, including evidence presented in this book. It seems to rely on its popularity for generating fashionable in-house controversies over inadequate materialist theories (for example, is religion a survival advantage or a glitch?), not on producing new insights.

As the classical philosophers understood, a divine Mind underlying the universe is the best explanation for the existence of the human soul. A science establishment that cannot tolerate that idea is poorly adapted to the reality it attempts to explore and hopes to explain.

CHAPTER 13

Does AI Really Change Everything? Anything?

SOME OF WHAT WE cover in this book—particularly the medical knowledge—is fairly new. But much of it, especially the philosophy, has been known for thousands of years. The overall picture points to the immortality of the human mind or soul.

However, a new question has arisen in the past few decades. Many technology experts think that advanced computers will soon become conscious, thinking like humans. But if programmers can replicate the human mind in silicon, wires, and electricity, how can that mind also be immortal? Is the computer mind immortal then too? Can humans *create* immortality as opposed to simply passing it on in our offspring? Whether programmers can really replicate the human mind in a computer is a critical question.

What's Behind the "Conscious Computer" Claims?

In June 2022, Google software engineer Blake Lemoine made headlines, claiming that LaMDA, a chatbot he was working on, had become sentient and might even have emotions.[1] Google denied the claim and fired Lemoine mid-uproar in July.[2]

But Lemoine's claim did not seem far-fetched to everyone. Philosopher Michael LaBossiere agrees with the basic idea: "After all, if a lump of organic goo can somehow think, then it is no more odd to think that a mass of circuitry or artificial goo could think." He added, "For those who think a soul is required to think, it is also no more bizarre for a ghost to be in a metal shell than in a meat shell."[3]

The idea has been percolating in neuroscience too. In 2010, South African neuroscientist Henry Markram, head of the Blue Brain Project, was pretty confident that conscious computers were not far off. A model of the human brain, like the one he was working on, would certainly help. He predicted, "Technologically, in terms of computers and techniques to acquire data, it will be possible to build a model of the human brain within 10 years."[4]

Of course, if human consciousness is simply an activity of the brain, a conscious computer may be only a matter of time. Or as Markram put it, "in the end there are going to be some very basic explanations for many things: emotions, awareness, consciousness, attention, perception, recognition."

Consciousness researcher Christof Koch was feeling pretty confident too in 2014, when he argued that machines would one day become self-aware. *Technology Review* summarized his approach: "Consciousness, he believes, is an intrinsic property of matter, just like mass or energy. Organize matter in just the right way, as in the mammalian brain, and *voilà*, you can feel."[5] Of course, after Koch lost a twenty-five-year wager with philosopher David Chalmers that a consciousness circuit would be found in the brain by 2023 (see chapter 9), he may have rethought that approach a bit. Certainly, his 2024 interview with Robert Lawrence Kuhn at *Closer to Truth* was much more thoughtful, less brash.[6]

But Koch's journey is not a trend. Many experts remain supremely confident that consciousness is an activity of the brain that can be duplicated in silicon. For example, a prominent research group, led by Koch's professional rival Stanislas Dehaene, spelled out in 2017: "Although centuries of philosophical dualism have led us to consider consciousness as unreducible to physical interactions, the empirical evidence is compatible with the possibility that consciousness arises from *nothing more than specific computations*"[7] (emphasis added). And therefore, the group concluded, machines could have consciousness. Along those lines, neuroscientist and tech whiz Ryota Kanai is hard at work trying to give computers "an introspective ability to report their internal mental states," because "such an ability is one of the main functions of consciousness."[8]

Some also think that computer intelligence would be an improvement on ours. Geoffrey Hinton, a former Google executive and professor at the University of Toronto and a 2024 Nobelist (Physics), told the *New Yorker* in 2023, "We should be concerned about digital intelligence taking over from biological intelligence" for that reason.[9]

Futurist and inventor Ray Kurzweil, who has made many correct predictions, is sure that there will be conscious computers as early as 2029: "We will get to the point—and this is not 100 years from now, it's like a few years—where they will act and respond just the way humans do. And if you say that they're not conscious, you'd have to say humans aren't conscious."[10]

Some techno-optimists avoid predicting the specifics of what will happen next. Sam Altman, the CEO of chatbot developer OpenAI and "a key player in the AI revolution" (*Time* magazine), pegs the revolution as really beginning in 2023. In 2024, he told *Time*, "we'll see way more of it, and by the time the end of this decade rolls around, I think the world is going to be in an unbelievably better place." But he also offers, "No one knows what happens next. I think the way technology goes, predictions are often wrong."[11]

Well, yes. Predictions *are* often wrong. To take but one example, at the 2016 Machine Learning and Market for Intelligence Conference in Toronto, Geoffrey Hinton predicted, "It's just completely obvious that within 5 years deep learning is going to do better than radiologists."[12] Yet in 2024, there was still a significant shortage of (human) radiologists in the United States.[13]

We also do not have Henry Markram's complete model of the human brain, predicted for 2020, either. We can't have it because, as *New Scientist* reported in 2024, "Strange new types of cells keep coming to light in the human brain. By the latest count, there are more than 3300, and we don't even know what most of them do."[14]

Are these unfulfilled predictions only a matter of unexpected delays that more research and better technology will solve?

Can Inanimate Objects Have Human Consciousness in Principle?

Durham University philosopher Philip Goff notes that panpsychism is an increasingly popular philosophical viewpoint. That's the view that everything has experience, consciousness is based on experience, and thus even inanimate things could become conscious:

> According to panpsychism, consciousness pervades the universe and is a fundamental feature of it. This doesn't mean that literally everything is conscious. The basic commitment is that the fundamental constituents of reality—perhaps electrons and quarks—have incredibly simple forms of experience, and the very complex experience of the human or animal brain is somehow derived from the experience of the brain's most basic parts.[15]

On that view, a sufficiently complex computer could become conscious in principle.[16]

So it comes down to two possibilities: If human consciousness has no spiritual component, perhaps machines that we build could become humanly conscious. But if human consciousness does have a spiritual component, advanced new technology is unlikely to supply it. Spiritual things are by nature immaterial and not subject to technology.

So then the question becomes, what does the weight of the evidence suggest? Are there aspects of AI that clearly signal that it is unlikely to become humanly conscious?

The Most Significant Limitation of Artificial Intelligence

A popular image of the mind–brain relationship is a computer. Even Wilder Penfield used that image at times. Some scientists go much further and double down, offering statements like: "The brain is not simply like a computer. It is literally a computer."[17] In their view, the mind is software and the brain is hardware.

But the comparison is off the mark. What would happen, for example, if you sawed a computer in half and threw one half away? Would the software work? Yet, as we have seen, the human mind, far from being a mere product of the brain, strives to work with whatever element of the brain is available, to maintain contact with the world.

The computer's most significant limitation is embedded in its name. The only thing it can do is *compute*. Programmer Eric Holloway explains what that entails: "There is no way to build a computer that cannot be reduced to the logic of 1's and 0's."[18] Computers that exhibit creativity or understand meaning are not possible because creativity and meaning cannot be reduced to 1's and 0's. That is true for any known form of AI. That limitation causes two obvious deficits: lack of creativity and lack of common sense.

Does AI Really Change Everything? Anything?

When There Is No Creativity...

University of Buenos Aires mathematician Gregory Chaitin is best known for identifying *Chaitin's unknowable number*. Briefly, the number exists and is a feature of computer programming, but it cannot be computed. As Baylor University computer engineering professor Robert J. Marks notes, Chaitin's work in identifying it is "an intellectually stunning piece of mathematics" with the clear philosophical and theological implication that our universe is not fully explainable in materialist terms.[19]

When Marks asked Chaitin whether he thought AI can be creative, he replied, "Creativity is what we *don't* know how to do. And so, it looks like it's a hard thing to program because, if we try to program productivity, that just becomes something mechanical and the frontier between what's creative and what isn't just moves a little forward."[20] In other words, anything that has been programmed and digitized is, by definition, already a solved problem, not an instance of ongoing creativity.

Fond of such paradoxes, he and Marks compare AI creativity to the *smallest uninteresting number*. Because such a number's very existence would make it interesting, it doesn't and can't exist.[21]

In his recent book, *Non-Computable You* (2022), Marks notes that the techno-optimists try to get around this problem by arguing (or hinting) that as AI gets more complex, human-like intelligence—and therefore, creativity—will somehow emerge. As a lifelong engineer, he offers an anecdote in response:

> Such unfounded optimism is akin to that of a naive young boy standing in front of a large pile of horse manure. He becomes excited and begins digging into the pile, flinging handfuls of manure over his shoulders.

The Immortal Mind

"With all this horse poop," he says, "there must be a pony in here somewhere!"[22]

But no, Dr. Marks says, there is no pony in the manure and no creativity or consciousness in the code: "All computer code is the result of human creativity—the written code itself can never be a source of creativity itself. The computer will perform as it is instructed by the programmer."[23]

Statistician Jeffrey Lee Funk and Pomona College business professor Gary N. Smith offered a striking example of AI's limitations where creativity is concerned at *MarketWatch*. Famously, in the game called Go, where the rules and the goal are clear, AI can now beat the best human players:

> If, however, the rules were changed or the goals could not be quantified, the algorithms would flop. If, for example, Go's 19-by-19 grid used today was changed to the 17-by-17 board that was used centuries ago, human experts would still play expertly but AI algorithms trained on a 19-by-19 board would be helpless. If the goal was to create an aesthetically pleasing pattern of stones, AI algorithms would be clueless.[24]

The fact that all AI can do is compute 1's and 0's raises other deep issues as well.

When There Is No Common Sense...

If the hallmark of AI is computation, a hallmark of the human mind is common sense—the things we just somehow know. Consider the headline "Dr. Gonzalez Gives Talk on Moon." Common sense tells us which of two possible interpretations of the headline is the writer's intent.[25]

Teaching AI common sense has long been the hope of researchers.[26] The problem they face is this: AI can simulate common sense but can't *understand* anything, which means that the common sense can never be real. As Dr. Marks puts it, "Computers can add the numbers 43 and 13 but do not understand what these numbers mean."[27]

Ironically, AI systems' indifference to meaning is precisely what makes computers so useful. As Gary Smith says, they can "manipulate words in many useful ways—e.g., spellchecking, searching, alphabetizing—without any understanding of the words they are manipulating. To know what words mean, they would have to understand the world we live in. They don't."[28]

Are the New Chatbots an Exception to All This?

Chatbots like ChatGPT-5 or Gemini (called large language models, or LLMs, in the trade) sound so much like conversation partners that people use them to help compose documents. If the software doesn't think, what is it doing?

Eric Holloway explains that chatbots like LaMDA, Gemini, and ChatGPT have been trained, using vast amounts of data, to respond automatically to *sequences of words* in human languages. This is similar to the "predictive text" option on your cell phone. It goes something like this: "Word *Smith* appears in the context of words *Dr.* and *Jane* 90 percent of the time, so with *Smith* and *Dr.*, predict *Jane*." Chatbots are more complicated than phones but they do roughly the same thing, based on the code they're given. They're not thinking about what you are saying and then responding; they are computing based on resources available on the internet.

Much of their predictive programming is done by the poorly paid human "grunts" worldwide who spend countless hours writing predictions.[29] As Krystal Kauffman, an organizer for such gig tech workers puts it, "You have all of these smart devices that people think are just magically smart; you have AI that people think magically appeared, and it is people like me [who make them work], and we're spread out all over the world."[30]

The chatbots themselves do not know what they are saying, any more than a thermometer knows what temperature it is or a watch knows what time it is. And complexity is irrelevant here—a state-of-the-art smartphone or chatbot is just as ignorant as an older one. They are tools—often very clever ones—that humans create.

It may seem strange, but chatbots can't even do math on their own. In 2022, Dr. Marks tested ChatGPT-3. The program failed elementary algebra because that was not part of its programming. Yes, a chatbot *can* be programmed to give correct answers to very complex algebra questions. But it hadn't been. So it didn't. And, absent any specific programming, it didn't know that it didn't know either.[31]

Similarly, in 2024, Gary Smith found that, when tested on financial planning questions involving arithmetic, Microsoft's Bing, OpenAI's GPT-4, and Google's Bard gave responses that were "seemingly authoritative but riddled with arithmetic and critical-thinking mistakes."[32] Again, that's because the program is not doing any thinking and cannot know when it isn't making sense. It is simply computing responses according to predictions. On that account, Dr. Smith does not think that financial planners will be replaced by LLMs anytime soon.

When Chatbots Hallucinate

Dr. Smith also wrote recently about an even stranger problem: *hallucination*. That is, chatbots may generate reams of wildly off-base material. Take ChatGPT's claim, which he witnessed, that the Soviets had sent forty-nine bears into space since 1957, bears with names such as "Alyosha," "Ugolek," "Zvezdochka," "Strelka," "Belka," "Pushinka," and "Vladimir." When references were requested, nonexistent links were provided. Microsoft's Copilot (formerly Bing) did not do any better when asked about bears in space.[33]

If chatbots aren't thinking, how can they make stuff up? At *Bulletin of the Atomic Scientists*, Sara Goudarzi points out that the bots merely provide programmed responses drawn from huge masses of information, correct or otherwise, not from thought processes: "Hallucination occurs because AI generative tools are trained by taking large amounts of web data and, left to themselves, treating all that information equally. To a large language model, information from a reliable encyclopedia could carry the same weight as a person's opinion expressed on a conspiracy blog."[34]

Is this problem fixable? Well, yes and no. Smith comments, "At some point, human handlers will train Copilot and other LLMs to respond that no bears have been sent into space but many thousands of other misstatements will fly under their radar. LLMs can generate falsehoods faster than humans can correct them."[35]

So while the specific claim about Soviet space bears can certainly be fixed, the underlying deficiencies of computation-only intelligence cannot. Generally, whatever human intelligence a chatbot appears to show is the melded output of the internet, programmers like Marks, crowdworkers like Lemoine, and grunts like Kauffman. What the chatbot adds is lightning-swift computation.

When Chatbots Simply Collapse into Nonsense

The chatbots depend on new human input. When that input declines, output starts to decay. Then, as Ben Lutkevitch explains at TechTarget, "probable events are overestimated and improbable events are underestimated...Over generations, models compound errors and more drastically misinterpret data."[36] In practice, it can look something like this:

> One recent study, published on the pre-print arXiv server, used a language model called OPT-125m to generate text about English architecture. After training the AI on that synthetic test over and over again, the 10th model's response was completely nonsensical and full of a strange obsession with jackrabbits.[37]

The only long-term remedy is more human input (including that of programmers, crowdworkers, and grunts), whether or not the tech industry likes paying for it. In short, however sophisticated AI becomes, a conscious, human-like mind is not built on 1's and 0's alone.

"Conscious" AI Makes No Logical Sense

Perhaps the most compelling reason that computers can never be conscious is that conscious computers make no *logical* sense. The hallmark of human consciousness is that every thought has a *meaning*. Every thought has an "aboutness"—that is, every thought is about something. You might think about lunch, or about your dog, or about mathematics.

In contrast, all computation is the matching of an input to an output according to a set of rules (an algorithm). The input, algorithm, and output of a computation are always mechanical in some way—a particular pattern of atoms or of 1's and 0's or of electrons on a screen. The hallmark of computation is that the process has no meaning in itself.

As Gary Smith notes, this limitation is necessary if the computer program is to be useful. Consider a word processing program. A student can use the program to type an essay arguing in favor of a viewpoint and can then use the same program to type an essay arguing against the same viewpoint. The computer doesn't "care" what the student types—computation is inherently blind to meaning. That is, in fact, what makes computation so useful. It is enormously flexible—you don't need a different word processing program for each opinion you want to express!

In fact, not only is AI incapable of consciousness, it is the opposite of consciousness. Computation is a blank slate on which we—who are conscious—express ourselves. It is a tool we use to communicate with others, to work more efficiently, to aid us in calculating, and so on.

Realizing these facts also helps us understand what is wrong with the view that the brain is a computer. Neuroscientists can and do *model* brain activity using computational models, but that does not mean that the brain itself is a computer. Nature provides answers that are framed by the questions we ask. If we model the brain as if it were a computer, it will seem like a computer. Many things can be studied using computational models—galaxies and living things and atoms—but that does not mean that galaxies and living things and atoms are computers. Computational models of the brain are useful in studying how the brain works, but the brain itself is not a computer, and the mind is not a kind of computation.

The Human Mind Does Require a Soul

Science writer John Horgan summed up the case against conscious computers at *Scientific American*: "Put bluntly: all evidence suggests that human and machine intelligence are radically different. And yet the myth of inevitability persists."[38] The myth persists because those who promote it naturally attract more publicity than skeptics do. But that does not mean that the promoters are more likely to prove correct. And if programmers have no idea how to make the most powerful machines do some of the things that humans easily do, the most likely reason is that some part of the human mind is, far from being a machine, immaterial.

Responding to philosopher Michael LaBossiere, we might say that thinking like a human being *does* require a soul. The AI project mainly shows that computing cannot manufacture souls—that is, human-like minds.

CONCLUSION

The Truths That Matter Most

MY CAREER AS A neurosurgeon and a scientist has been a voyage. I set out to heal, of course, but I also set out to explore, to learn the truth about who we are, and why we are here, and where we are going. I needed my experience as a neurosurgeon, as well as neuroscience, philosophy, and theology, to show me the way. I also needed prayer, because these are questions that we can't answer unaided.

All I have found points to one truth—we have spiritual immortal souls. I see this in my operating room, in my clinics, in the remarkable research of world-class neuroscientists, in the insights of ancient and modern philosophers and theologians, and in my own moments of quiet reflection and prayer. This is a gratifying insight, for sure, but another question emerges, the question that led me to write this book: Why does all this matter? Why is the fact that we have spiritual immortal souls important? Should we live our lives differently because of it?

Of course we should. The fact that we have souls, that our souls are spiritual and made in God's image and are destined for immortal life, is the most important thing about us. Nothing matters more.

Conclusion

Marriage and Family

The spirituality and immortality of our souls is central to our marriages and our children. Marriage is the union of two immortal spirits, not just the union of two bodies. Spirits like us do not live in time, so marriage is an eternal bond of mutual love and care. In Catholicism, marriage is a sacrament, a temporal realization of an eternal reality. Our spiritual nature in marriage is expressed clearly in our most astonishing act—our participation in the creation of new life in our children.

Our children have spiritual immortal souls just as we do, and by God's grace, they are created *through* us. Mother and father supply the biological necessities, but God directly creates the soul, because only Spirit can give rise to spirit. This is an immeasurable blessing, and a daunting responsibility. With each conception, a new spiritual soul enters the world in our care. Perhaps more than anyone, our children carry with them the eternal blessings, love, and trials that we give them. Parenthood is an awesome responsibility with eternal ramifications.

Free Will

The truth about human souls is pivotal in the modern debate about whether we have free will. Neuroscience clearly points to the freedom of the will. The very nature of our souls—that we are *spiritual*—means that our choices are ultimately not determined by matter. We are of course influenced by our bodies—we are material bodies given life by spiritual souls—but we by nature have the ability to accept or refuse the urges of our bodies. Free will is real. We are naturally free and responsible, by virtue of our spiritual souls.

Conclusion

Human Dignity

The spiritual nature of our souls has enormous implications for our treatment of our physically handicapped brothers and sisters. Each person, no matter how disabled, is an immortal spiritual creature. People who are physically handicapped are endowed with spiritual souls just as we are, and they are deserving of respect, love, and care. Our elderly loved ones in the throes of dementia retain their spiritual nature, even when it is crippled by Alzheimer's disease or other brain disorders that may afflict us in our final years in this life. In the Christian tradition, the disabled warrant special protection and care, as they are so often the least among us. No matter how crippling physical disabilities may be, the most horrific disabilities with which we, as immortal creatures, struggle are spiritual, not physical.

Respect for young human lives in the womb takes on new meaning and urgency in light of the spirituality and immortality of the human soul. It is a scientific fact that human life begins at conception and continues to natural death. Human life is ensoulment—the soul is just the difference between a live human being and a dead human being—so the child in the womb has an immortal soul from the moment of conception. Unborn children are not parts of the mother's body, or mere "tissue" to be disposed of if unwanted or inconvenient. Children in the womb have the dignity appropriate to all human beings—they are immortal creatures created in God's image.

The same dignity of every human being in the womb is found in every human being of every race. There are no "White" or "Black" human souls, no "Asian" souls, no "Hispanic" souls of greater or lesser dignity on account of racial or ethnic heritage. Our bodies—the manifestation of human racial differences—will disintegrate at death, but our spiritual souls will live on in eternity. We are all one in spirit.

Conclusion

Likewise, men and women alike have equal dignity appropriate to them by virtue of their spiritual souls. Sex is created at conception, not assigned at birth, and medical or surgical mutilation of the body do not and cannot change sexual or spiritual reality. No person—no spiritual soul—is created in the wrong body. Our brothers and sisters suffering from gender dysphoria need our compassion, love, and help, which, in these times, too often entails protecting them from medical and surgical mutilation of their God-given bodies.

War

War takes on a new meaning when we realize that every human being—including the "enemy" we are too often asked to hate—has a spiritual and immortal nature just as we do. Even a modicum of respect for human dignity precludes aggressive war and especially the targeting of innocents. As Aleksandr Solzhenitsyn pointed out, the line between good and evil runs not between nations or races or religions but through every human heart. Our worst enemy is not other people. It is in ourselves and in the evil that prowls about the world seeking our eternal damnation. The real war we fight is spiritual.

Human Rights

The reality of our immortal spiritual souls orders our society in vitally important ways. The truth about the human soul is crucial for our rights and liberties. The American Declaration of Independence specifically invokes the equality of all human beings and holds that our rights are granted to us by our Creator:

> We hold these truths to be self-evident, that all men are created equal, that they are endowed by their Creator with

Conclusion

certain unalienable Rights, that among these are Life, Liberty and the pursuit of Happiness. That to secure these rights, Governments are instituted among Men, deriving their just powers from the consent of the governed.

Ascribing natural human rights to all people only makes sense for human beings with spiritual souls. Animals without an immortal spirit lack the dignity that would require that they be endowed with natural rights. In the view of Hannah Arendt, author of *The Origins of Totalitarianism* (1951), the Darwinian view of man as an evolved animal lacking spiritual and immortal dignity is the ideological basis for totalitarian systems of government:

> Underlying the Nazis' belief in race laws as the expression of the law of nature in man, is Darwin's idea of man as the product of a natural development which does not necessarily stop with the present species of human beings, just as under the Bolsheviks' belief in class-struggle as the expression of the law of history lies Marx's notion of society as the product of a gigantic historical movement which races according to its own law of motion to the end of historical times when it will abolish itself.[1]

Totalitarians such as Nazis and Communists saw men not as fellow human beings with immortal spiritual souls but as human livestock subject to racial, historical, and economic laws of nature. The totalitarian movement strove to quicken the law of natural selection, to cull the weak and undesirable "species" so as to herald a new version of evolved mankind.

The totalitarian view of man is explicitly Darwinian. In his funeral oration for Karl Marx, Friedrich Engels openly credited Darwin for Marxist ideology: "Just as Darwin discovered the law of development of organic nature, so Marx discovered the law of development of human history."[2]

Totalitarians have always sought to exterminate religion and suppress the spiritual understanding of human beings—they understand that society's best protection from totalitarianism is reverence for God and acknowledgment of the fallibility and need for salvation of human beings with immortal spiritual souls.

Science

The spirituality and immortality of our souls point to additional consequences for those of us who do science and who care about science. As we have seen in these pages, neuroscience for over a century has clearly pointed to the existence of the human soul, to its spirituality and to its immortality. Yet these obvious scientific insights are rarely mentioned by scientists, either in scientific publications or in their statements as public figures. This is a scandal. Some leading and courageous neuroscientists—Wilder Penfield is a prime example—have told the truth about their science. But most remain silent about the human soul, its spiritual nature and its immortality, even though their research unequivocally points to these truths. The silence about the spiritual nature and the immortality of the human soul in the science community is a lie by omission.

The truth about our souls has enormous implications for other branches of science as well. As we have noted in these pages, our souls cannot have evolved by Darwinian evolution, which only has creative power (if it has any creative power at all) over matter, not spirit. Man's soul did not evolve from apes, or from any material creature, because the spiritual human soul is not the kind of thing an ape has, nor is it the kind of thing that can come about by natural selection. Only Spirit gives rise to spirit. We are created, not evolved.

Conclusion

Dying Without the Truth

It is not only important that we personally recognize the truth about the human soul—we should help others to learn it as well. To die in ignorance of the philosophical, theological, and scientific truth that we have spiritual immortal souls is always a human tragedy.

William Provine was the Andrew H. and James S. Tisch Distinguished University Professor at Cornell University. He held appointments as a professor in the Departments of History, Science and Technology Studies, and Ecology and Evolutionary Biology. Despite his brilliance, he was an atheist and denied the immortality and spirituality of the soul. In 1994, he said:

> Let me summarize my views on what modern evolutionary biology tells us loud and clear—and these are basically Darwin's views. There are no gods, no purposes, no goal-directed forces of any kind. There is no life after death. When I die, I am absolutely certain that I am going to be dead. That's the end for me. There is no ultimate foundation for ethics, no ultimate meaning to life, and no free will for humans, either.[3]

Provine died in 2015, still refusing to accept the philosophical, theological, and scientific truth that his soul is immortal. To die in such ignorance and falsehood is a horror and a tragedy.

What Matters

Finally, what does this truth about our immortal spiritual soul matter to us personally? This truth, as we have seen, is the most important thing about us. Because we are immortal, *everything* we do matters.

Conclusion

How we go about our daily lives leaves eternal tracks, and every person we meet is as immortal as we are, destined to live in eternity with the influence we have on them and the memories they carry. Every kind act and cruel word reverberates in eternity.

This gives me pause. I see that what so many people who have near-death experiences report is true—that at our death, we are shown everything in our lives, in real time. Nothing is ever lost, for better or worse. None of us—and certainly not I—could stand before a review of every moment of my life in peace and equanimity. I have much to answer for, in what I have done and what I have failed to do. And what I have done lasts forever.

As frightening as this prospect is, it opens the door to a deep hope. We cannot enter eternity justified by our own lives—I have not lived a single day that would make me proud before God. The immortality of my soul and the eternal consequences of my life mean that I need to be forgiven, I need to place my trust in Someone who bore my sins, and I need to accept His grace. Scripture, self-reflection, and science all point to this reality about us.

We are embodied spirits who need God's love and forgiveness, now and in eternity.

Acknowledgments

We would like to acknowledge the invaluable contributions of our agent, Giles Anderson, and John West, vice president of the Discovery Institute.

We very much appreciate the hard work and dedication of Beth Adams, Hailey Juen, Roland Ottewell, Stacey Sharp, Catherine Hoort, Katie Robison, Taylor Peterson, Patsy Jones, Ryan Peterson, Abigail Skinner, and others on our book project.

We are grateful to our many advance readers, including Mario Beauregard, Sharon Dirckx, Sarah Handler, Fr. Martin Hilbert, OC, Fr. Lee Kenyon, Nancy Pearcey, Scott Ventureyra, Stephanie West Allen, and many others for their insights and encouragement.

Michael Egnor would like to thank his patients for their courage and perseverance in the face of life-altering brain injuries. They exemplify the nobility of the human soul. He would also like to thank Professor Edward Feser for his brilliant introductions to Thomistic psychology.

Denyse O'Leary would like to thank computer technician George Cuppage for his savvy when the machine seemed, at times, like an enemy force.

Notes

Chapter 1: The Brain Can Be Split but Not the Mind

1 V. Siffredi et al., "Structural Neuroplastic Responses Preserve Functional Connectivity and Neurobehavioural Outcomes in Children Born Without Corpus Callosum," *Cerebral Cortex* 31, no. 2 (2021): 1227–39. doi.org/10.1093/cercor/bhaa289.

2 C. M. Kaculini, A. J. Tate-Looney, and A. Seifi, "The History of Epilepsy: From Ancient Mystery to Modern Misconception," *Cureus* 13, no. 3 (March 17, 2021): e13953. doi: 10.7759/cureus.13953. PMID: 33880289; PMCID: PMC8051941.

3 As given and discussed here: R. S. Fisher et al., "Epileptic Seizures and Epilepsy: Definitions Proposed by the International League Against Epilepsy (ILAE) and the International Bureau for Epilepsy (IBE)," *Epilepsia* 46, no. 4 (April 2005): 470–72. doi:10.1111/j.0013-9580.2005.66104.x. PMID: 15816939.

4 E. Magiorkinis et al., "Highights in the History of Epilepsy: The Last 200 Years," *Epilepsy Research and Treatment* (2014), 2014:582039. doi:10.1155/2014/582039. Epub 2014 Aug 24. PMID: 25210626; PMCID: PMC4158257.

5 Francis Crick, introduction to *The Astonishing Hypothesis: The Scientific Search for Soul* (New York: Scribner, 1994).

6 For example, R. Joseph, "Dual Mental Functioning in a Split-Brain Patient," *Journal of Clinical Psychology* 44, no. 5 (September 1988): 770–79. doi:10.1002/1097-4679(198809)44:5<770:aid-jclp2270440518>3.0.co;2-5. PMID: 3192716.

7 Alyssa Anderson, "What Is Alien Hand Syndrome?," WebMD, April 8, 2022, www.webmd.com/brain/what-is-alien-hand-syndrome. Accessed May 23, 2023.

8 J. Sergent, "Unified Response to Bilateral Hemispheric Stimulation by a Split-Brain Patient," *Nature* 305, no. 5937 (October 27–November 2, 1983): 800–802. doi:10.1038/305800a0. PMID: 6633650.

9 M. C. Corballis and J. Sergent, "Judgements About Numerosity by a Commissurotomized Subject," *Neuropsychologia* 30, no. 10 (October 1992): 865–76. doi:10.1016/0028-3932(92)90032-h. PMID: 1436434.

Notes

10 J. Sergent, "A New Look at the Human Split Brain," *Brain* 110, Pt 5 (October 1987): 1375–92. doi:10.1093/brain/110.5.1375. PMID: 3676706.
11 Y. Pinto, E. H. F. de Haan, and V. A. F. Lamme, "The Split-Brain Phenomenon Revisited: A Single Conscious Agent with Split Perception," *Trends in Cognitive Sciences* 21, no. 11 (November 2017): 835–51. doi:10.1016/j.tics.2017.09.003. Epub 2017 Sep 25. PMID: 28958646.
12 L. D. Ladino, S. Rizvi, and J. F. Téllez-Zenteno, "The Montreal Procedure: The Legacy of the Great Wilder Penfield," *Epilepsy & Behavior* 83 (June 2018): 151–61. doi:10.1016/j.yebeh.2018.04.001. Epub 2018 Apr 26. PMID: 29705626.
13 Ladino et al., "The Montreal Procedure."
14 Ladino et al., "The Montreal Procedure."
15 "Electrocorticography," ScienceDirect, www.sciencedirect.com/topics/neuroscience/electrocorticography. Accessed September 22, 2023.
16 "Electrocorticography," ScienceDirect.
17 "Penfield," The Neuro, www.mcgill.ca/neuro/about/history/wilder-graves-penfield. Accessed September 21, 2023. A reenactment of the 1934 surgery is available at Historica Canada, "Heritage Minutes: Wilder Penfield," YouTube, www.youtube.com/watch?v=pUOG2g4hj8s. Accessed May 6, 2024.
18 Wilder Penfield, *The Mystery of the Mind* (Princeton University Press, 1975), 12–13.
19 "Electrocorticography," ScienceDirect.
20 Work on the homunculus has since been updated. See E. M. Gordon et al., "A Somato-cognitive Action Network Alternates with Effector Regions in Motor Cortex," *Nature* 617, no. 7960 (May 2023): 351–59. doi:10.1038/s41586-023-05964-2. Epub 2023 Apr 19. PMID: 37076628; PMCID: PMC10172144.
21 Penfield, *Mystery*, 46.
22 Ladino et al., "The Montreal Procedure," 151–61.
23 Penfield, *Mystery*, 55.
24 Penfield, *Mystery*, 27.
25 Penfield, *Mystery*, 88.
26 Wilder Penfield and Joseph Evans, "The Frontal Lobe in Man: A Clinical Study of Maximum Removals," *Brain* 58, Issue 1 (March 1935): 115–33. doi.org/10.1093/brain/58.1.115.
27 Vaughan Bell, "The Hardest Cut: Penfield and the Fight for His Sister," Mind Hacks, June 28, 2007, https://mindhacks.com/2007/06/28/the-hardest-cut-penfield-and-the-fight-for-his-sister/. Accessed December 12, 2023.
28 Y. J. Cho et al., "Simple Partial Status of Forced Thinking Originated in the Mesial Temporal Region: Intracranial Foramen Ovale Electrode Recording and Ictal PET," *Journal of Epilepsy Research* 1, no. 2 (December 30, 2011): 77–80. doi:10.14581/jer.11015. PMID: 24649451; PMCID: PMC3952329; G. Maggu et al., "Behavioral Presentations of Focal Onset Seizures: A Case Series," *Industrial Psychiatry Journal* 30, Suppl. 1 (October 2021): S204–9. doi:10.4103/0972-6748.328869. Epub 2021 Oct 22. PMID: 34908691; PMCID: PMC8611604.
29 Michael Egnor, "Do Forced Thinking Seizures Show That Abstract Thought Is a Material Thing?," *Mind Matters News*, July 2, 2019, https://mindmatters.ai/2019/07/do-forced-thinking-seizures-show-that-abstract-thought-is-a-material-thing/. Accessed May 6, 2024.

Notes

30. M. F. Mendez, M. M. Cherrier, and K. M. Perryman, "Epileptic Forced Thinking from Left Frontal Lesions," *Neurology* 47, no. 1 (July 1996): 79–83. doi:10.1212/wnl.47.1.79. PMID: 8710129.
31. Penfield, *Mystery*, 31. All page number citations not otherwise noted below are from this source.
32. Penfield, *Mystery*, xxv.
33. Penfield, *Mystery*, xviii–xix.
34. Penfield, *Mystery*, 85.
35. Penfield, *Mystery*, xxi.

Chapter 2: How Much Brain Does the Mind Need?

1. Michael Egnor, "A Map of the Soul," *First Things*, June 29, 2017, www.firstthings.com/web-exclusives/2017/06/a-map-of-the-soul. Accessed May 6, 2024.
2. Emily Mae Czachor, "Emilia Clarke Reflects on 'Remarkable' Ability to Speak Despite Losing 'Quite a Bit' of Her Brain to Aneurysms," CBS News, July 19, 2022, www.cbsnews.com/news/emilia-clarke-brain-aneurysms-remarkable-recovery-ability-to-speak/. Accessed May 6, 2024.
3. Jon Hamilton, "Meet the 'Glass-Half-Full Girl' Whose Brain Rewired After Losing a Hemisphere," NPR, March 22, 2023, www.npr.org/sections/health-shots/2023/03/22/1165131907/neuroplasticity-plasticity-glass-half-full-girl. Accessed May 6, 2024.
4. Laura Sanders, "Some People with Half a Brain Have Extra Strong Neural Connections," *ScienceNews*, November 19, 2019, www.sciencenews.org/article/some-people-with-half-brain-have-extra-strong-neural-connections. Accessed May 6, 2024; D. Kliemann et al., "Intrinsic Functional Connectivity of the Brain in Adults with a Single Cerebral Hemisphere," *Cell Reports* 29, no. 8 (November 2019): 2398–407.e4. doi:10.1016/j.celrep.2019.10.067. PMID: 31747608; PMCID: PMC6914265.
5. Bob Yirka, "Adults Who, as Children, Had Half Their Brain Removed Still Able to Score Well with Face and Word Recognition," MedicalXpress, August 17, 2022, https://medicalxpress.com/news/2022-08-adults-children-brain-score-word.html. Accessed May 6, 2024; M. C. Granovetter et al., "With Childhood Hemispherectomy, One Hemisphere Can Support—but Is Suboptimal for—Word and Face Recognition," *Proceedings of the National Academy of Sciences of the United States of America* 119, no. 44 (November 2022): e2212936119. doi:10.1073/pnas.2212936119. Epub 2022 Oct 25. PMID: 36282918; PMCID: PMC9636967.
6. T. T. Liu et al., "Successful Reorganization of Category-Selective Visual Cortex Following Occipito-temporal Lobectomy in Childhood," *Cell Reports* 24, no. 5 (July 31, 2018): 1113–22.e6. doi:10.1016/j.celrep.2018.06.099. PMID: 30067969; PMCID: PMC6152879.
7. Michael Egnor, "Science and the Soul," *The Plough*, August 20, 2018, www.plough.com/en/topics/justice/reconciliation/science-and-the-soul. Accessed May 6, 2024.
8. Université de Genève, "A Malformation Illustrates the Incredible Plasticity of the Brain," *ScienceDaily*, October 30, 2020, www.sciencedaily.com/releases/2020/10/201030122550.htm. Accessed May 6, 2024; V. Siffredi et al., "Structural Neuroplastic Responses Preserve Functional Connectivity and Neurobehavioural

Notes

Outcomes in Children Born Without Corpus Callosum," *Cerebral Cortex* 31, no. 2 (January 5, 2022): 1227–39. doi:10.1093/cercor/bhaa289. PMID: 33108795.
9 Helen Santoro, "The Curious Hole in My Head," *New York Times*, September 4, 2022, www.nytimes.com/2022/09/04/science/brain-language-research.html. Accessed May 6, 2024.
10 Santoro, "The Curious Hole."
11 Helen Thomson, "Woman of 24 Found to Have No Cerebellum in Her Brain," *New Scientist*, September 10, 2014, www.newscientist.com/article/mg22329861-900-woman-of-24-found-to-have-no-cerebellum-in-her-brain/. Accessed May 6, 2024.
12 Tom Stafford, "Can You Live a Normal Life with Half a Brain?," BBC, December 17, 2014, www.bbc.com/future/article/20141216-can-you-live-with-half-a-brain. Accessed May 6, 2024.
13 F. Yu et al., "A New Case of Complete Primary Cerebellar Agenesis: Clinical and Imaging Findings in a Living Patient," *Brain* 138, Pt 6 (June 2015): e353. doi:10.1093/brain/awu239. Epub 2014 Aug 22. PMID: 25149410; PMCID: PMC4614135.
14 S. S. Asaridou et al., "Language Development and Brain Reorganization in a Child Born Without the Left Hemisphere," *Cortex* 127 (June 2020): 290–312. doi:10.1016/j.cortex.2020.02.006. Epub 2020 Feb 29. PMID: 32259667; PMCID: PMC8025291.
15 P. Pavone et al., "Hemihydranencephaly: Living with Half Brain Dysfunction," *Italian Journal of Pediatrics* 39 (January 16, 2013): 3. doi:10.1186/1824-7288-39-3. PMID: 23324549; PMCID: PMC3564735.
16 L Feuillet, H. Dufour, and J. Pelletier, "Brain of a White-Collar Worker," *Lancet* 370, no. 9583 (July 21, 2007): 262. doi:10.1016/S0140-6736(07)61127-1. PMID: 17658396.
17 "Man with Tiny Brain Shocks Doctors," *New Scientist*, July 20, 2007, www.newscientist.com/article/dn12301-man-with-tiny-brain-shocks-doctors/. Accessed May 6, 2024.
18 Stafford, "Can You Live a Normal Life with Half a Brain?"
19 See, for example, Stafford, "Can You Live a Normal Life with Half a Brain?"
20 Jose I. Sandoval and Orlando De Jesus, "Hydranencephaly," National Library of Medicine, Last Update: June 26, 2023, www.ncbi.nlm.nih.gov/books/NBK558991/. Accessed September 24, 2023.
21 "Diagnosis may be delayed for several months because early behavior appears to be relatively typical." From "Hydranencephaly," National Institute of Neurological Disorders and Stroke, www.ninds.nih.gov/health-information/disorders/hydranencephaly. Accessed September 25, 2023.
22 G. N. McAbee, A. Chan, and E. L. Erde, "Prolonged Survival with Hydranencephaly: Report of Two Patients and Literature Review," *Pediatric Neurology* 23, no. 1 (July 2000): 80–84. doi:10.1016/s0887-8994(00)00154-5. PMID: 10963978.
23 J. S. Bae, M. U. Jang, and S. S. Park, "Prolonged Survival to Adulthood of an Individual with Hydranencephaly," *Clinical Neurology and Neurosurgery* 110, no. 3 (March 2008): 307–9. doi:10.1016/j.clineuro.2007.12.003. Epub 2008 Jan 28. PMID: 18222607.
24 B. Merker, "Consciousness Without a Cerebral Cortex: A Challenge for Neuroscience and Medicine," *Behavioral and Brain Sciences* 30, no. 1 (February 2007): 63–81; discussion 81–134. doi:10.1017/S0140525X07000891. PMID: 17475053.
25 Bruce Bower, "Consciousness in the Raw," *Science News*, September 11, 2007, www.sciencenews.org/article/consciousness-raw. Accessed May 6, 2024.
26 Merker, "Consciousness Without a Cerebral Cortex."

Notes

27 D. A. Shewmon, G. L. Holmes, and P. A. Byrne, "Consciousness in Congenitally Decorticate Children: Developmental Vegetative State as Self-Fulfilling Prophecy," *Developmental Medicine & Child Neurology* 41, no. 6 (June 1999): 364–74. doi:10.1017/s0012162299000821. PMID: 10400170.

28 Ralph Lewis, "An Overview of the Leading Theories of Consciousness," *Psychology Today*, October 2, 2023, www.psychologytoday.com/us/blog/finding-purpose/202308/an-overview-of-the-leading-theories-of-consciousness. Accessed May 6, 2024.

29 "Consciousness: Is It in the Cerebral Cortex or the Brain Stem?," *Mind Matters News*, November 6, 2021, https://mindmatters.ai/2021/11/consciousness-is-it-in-the-cerebral-cortex-or-the-brain-stem/. Accessed May 6, 2024.

30 Thomas Nagel, "What Is It Like to Be a Bat?," *Philosophical Review* 83, no. 4 (October 1974): 435–50, www.academia.edu/40280185/Philosophical_Review_What_Is_It_Like_to_Be_a_Bat.

31 Tom Bartlett, "Has Consciousness Lost Its Mind?," *Chronicle of Higher Education*, June 6, 2018, www.chronicle.com/article/has-consciousness-lost-its-mind/. Accessed May 6, 2024.

32 M. Lenharo, "Consciousness Theory Slammed as 'Pseudoscience'—Sparking Uproar," *Nature*, September 20, 2023. doi:10.1038/d41586-023-02971-1. Epub ahead of print. PMID: 37730789.

Chapter 3: The Mind Is Hard to Just Put Out

1 S. Laureys et al. and the European Task Force on Disorders of Consciousness, "Unresponsive Wakefulness Syndrome: A New Name for the Vegetative State or Apallic Syndrome," *BMC Medicine* 8 (November 2010): 68. doi:10.1186/1741-7015-8-68. PMID: 21040571; PMCID: PMC2987895.

2 D. Cruse et al., "Detecting Awareness in the Vegetative State: Electroencephalographic Evidence for Attempted Movements to Command," *PLoS One* 7, no. 11 (2012): e49933. doi:10.1371/journal.pone.0049933. Epub 2012 Nov 21. PMID: 23185489; PMCID: PMC3503880.

3 P. S. Mueller, "The Terri Schiavo Saga: Ethical and Legal Aspects and Implications for Clinicians," *Polskie Archiwum Medycyny Wewnetrznej* 119, no. 9 (September 2009): 574–81. PMID: 19776703.

4 Mueller, "The Terri Schiavo Saga."

5 Richard Alleyne and Martin Beckford, "Patients in 'Vegetative' State Can Think and Communicate," *Telegraph*, February 4, 2010, www.telegraph.co.uk/news/health/news/7150119/Patients-in-vegetative-state-can-think-and-communicate.html. Accessed May 7, 2024.

6 M. M. Monti et al., "Willful Modulation of Brain Activity in Disorders of Consciousness," *New England Journal of Medicine* 362, no. 7 (February 18, 2010): 579–89. doi:10.1056/NEJMoa0905370. Epub 2010 Feb 3. PMID: 20130250.

7 Alleyne and Beckford, "Patients in 'Vegetative' State."

8 D. Cruse et al., "Bedside Detection of Awareness in the Vegetative State: A Cohort Study," *Lancet* 378, no. 9809 (December 17, 2011): 2088–94. doi:10.1016/S0140-6736(11)61224-5. Epub 2011 Nov 9. PMID: 22078855.

9 Jan Claassen and Brian L. Edlow, "Some People Who Appear to Be in a Coma May Actually Be Conscious," *Scientific American*, November 1, 2022, www.scientificamerican

Notes

.com/article/some-people-who-appear-to-be-in-a-coma-may-actually-be-conscious/. Accessed May 7, 2024; J. Claassen et al., "Detection of Brain Activation in Unresponsive Patients with Acute Brain Injury," *New England Journal of Medicine* 380, no. 26 (June 27, 2019): 2497–505. doi:10.1056/NEJMoa1812757. PMID: 31242361.

10. Claassen and Edlow, "Some People Who Appear."
11. Y. G. Bodien et al., "Cognitive Motor Dissociation in Disorders of Consciousness," *New England Journal of Medicine* 391, no. 7 (August 15, 2024): 598–608. doi:10.1056/NEJMoa2400645.
12. J. T. Giacino et al., "The Minimally Conscious State: Definition and Diagnostic Criteria," *Neurology* 58, no. 3 (February 12, 2002): 349–53. doi:10.1212/wnl.58.3.349. PMID: 11839831.
13. D. J. Strauss et al., "Life Expectancy of Children in Vegetative and Minimally Conscious States," *Pediatric Neurology* 23, no. 4 (2000): 1–8. CiteSeerX 10.1.1.511.2986. doi:10.1016/S0887-8994(00)00194-6. PMID 11068163.
14. Steven Novella, "Intelligent Design of the Brain," *Neurologicablog*, February 11, 2008, https://theness.com/neurologicablog/intelligent-design-of-the-brain/. Accessed May 7, 2024.
15. Michael Egnor, "Atheist Psychiatrist Misunderstands Evidence for an Immaterial Mind," *Mind Matters News*, June 30, 2019, https://mindmatters.ai/2019/06/atheist-psychiatrist-misunderstands-evidence-for-immaterial-mind/. Accessed May 7, 2024.
16. Alex Godfrey, "'The Clouds Cleared': What Terminal Lucidity Teaches Us About Life, Death and Dementia," *Guardian*, February 23, 2021, www.theguardian.com/society/2021/feb/23/the-clouds-cleared-what-terminal-lucidity-teaches-us-about-life-death-and-dementia. Accessed May 7, 2024.
17. Jordan Kinard, "Why Dying People Often Experience a Burst of Lucidity," *Scientific American*, June 12, 2023, www.scientificamerican.com/article/why-dying-people-often-experience-a-burst-of-lucidity/. Accessed May 7, 2024.
18. This may be the first use of the term. M. Nahm, "Terminal Lucidity in People with Mental Illness and Other Mental Disability: An Overview and Implications for Possible Explanatory Models," *Journal of Near-Death Studies* 28, no. 2 (Winter 2009): 87–106.
19. Michael Nahm, "Terminal Lucidity in People with Mental Illness and Other Mental Disability: An Overview and Implications for Possible Explanatory Models," *Journal of Near-Death Studies* 28, no. 2 (Winter 2009); Carlos S. Alvarado, "Neglected Near-Death Phenomena," *Journal of Near-Death Studies* 24, no. 3 (Spring 2006): 131–51.
20. Michigan Medicine—University of Michigan, "Moments of Clarity in Dementia Patients at End of Life: Glimmers of Hope?," *ScienceDaily*, June 28, 2019, www.sciencedaily.com/releases/2019/06/190628182305.htm.
21. Jesse Bering, "One Last Goodbye: The Strange Case of Terminal Lucidity," *Scientific American*, November 25, 2014, https://blogs.scientificamerican.com/bering-in-mind/one-last-goodbye-the-strange-case-of-terminal-lucidity/. Accessed May 7, 2024.
22. Neuroskeptic [pen name], "Terminal Lucidity: Myth, Mystery or Miracle?," *Discover*, August 9, 2014, www.discovermagazine.com/mind/terminal-lucidity-myth-mystery-or-miracle. Accessed May 7, 2024.
23. Caitlin Geng, "What to Know About Terminal Lucidity and Dementia," *Medical News Today*, August 30, 2022, www.medicalnewstoday.com/articles/terminal-lucidity-dementia. Accessed May 7, 2024.

Notes

24 S. Parnia et al., "AWAreness during REsuscitation—II: A Multi-Center Study of Consciousness and Awareness in Cardiac Arrest," *Resuscitation* (July 7, 2023): 109903. doi: 10.1016/j.resuscitation.2023.109903. Epub ahead of print. PMID: 37423492.
25 Kinard, "Why Dying People."
26 Elsevier, "New Evidence Indicates Patients Recall Death Experiences After Cardiac Arrest," *ScienceDaily*, September 14, 2023, www.sciencedaily.com/releases/2023/09/230914175140.htm; S. Parnia S et al., "AWAreness during REsuscitation—II."
27 D. B. Ney, A. Peterson, and J. Karlawish, "The Ethical Implications of Paradoxical Lucidity in Persons with Dementia," *Journal of the American Geriatrics Society* 69, no. 12 (December 2021): 3617–22. doi:10.1111/jgs.17484. Epub 2021 Oct 10. PMID: 34628640; PMCID: PMC9924090.
28 Kinard, "Why Dying People."
29 Will Cairns, "Terminal Lucidity: When Dying People Wake Up," *InSight+*, March 15, 2021, https://insightplus.mja.com.au/2021/8/terminal-lucidity-when-dying-people-wake-up/. Accessed May 7, 2024.
30 Cairns, "Terminal Lucidity."
31 M. Nahm and B. Greyson, "Terminal Lucidity in Patients with Chronic Schizophrenia and Dementia: A Survey of the Literature," *Journal of Nervous and Mental Disease* 197, no. 12 (December 2009): 942–44. doi:10.1097/NMD.0b013e3181c22583. PMID: 20010032.
32 A. Peterson et al., "What Is Paradoxical Lucidity? The Answer Begins with Its Definition," *Alzheimer's & Dementia* 18, no. 3 (March 2022): 513–21. doi:10.1002/alz.12424. Epub 2021 Aug 2. PMID: 34338400; PMCID: PMC8807788.
33 A. Batthyány and B. Greyson, "Spontaneous Remission of Dementia Before Death: Results from a Study on Paradoxical Lucidity," *Psychology of Consciousness: Theory, Research, and Practice* 8, no. 1 (2020): 1–8. doi.org/10.1037/cns0000259.
34 Peterson et al., "What Is Paradoxical Lucidity?"
35 Peterson et al., "What Is Paradoxical Lucidity?"
36 "2023 Alzheimer's Disease Facts and Figures," *Alzheimer's & Dementia* 19, no. 4 (April 2023): 1598–1695. doi:10.1002/alz.13016. Epub 2023 Mar 14.PMID: 36918389.
37 Peterson et al., "What Is Paradoxical Lucidity?"
38 S. G. Post, "'Is Grandma Still There?' A Pastoral and Ethical Reflection on the Soul and Continuing Self-Identity in Deeply Forgetful People," *Journal of Pastoral Care & Counseling* 70, no. 2 (June 2016): 148–53. doi:10.1177/1542305016644739. PMID: 27281763.
39 T. F. Brady et al., "Visual Long-Term Memory Has a Massive Storage Capacity for Object Details," *Proceedings of the National Academy of Sciences of the United States of America* 105, no. 38 (September 23, 2008): 14325–29. doi:10.1073/pnas.0803390105. Epub 2008 Sep 11. PMID: 18787113; PMCID: PMC2533687.
40 Roger Lewin, "Is Your Brain Really Necessary?," *Science* 210 (1980): 1232–34. doi: 10.1126/science.7434023.
41 L. Feuillet, H. Dufour, and J. Pelletier, "Brain of a White-Collar Worker," *Lancet* 370, no. 9583 (July 21, 2007): 262. doi:10.1016/S0140-6736(07)61127-1. PMID: 17658396; D. R. Forsdyke, "Wittgenstein's Certainty Is Uncertain: Brain Scans of Cured Hydrocephalics Challenge Cherished Assumptions," *Biological Theory* 10 (2015): 336–42. doi.org/10.1007/s13752-015-0219-x.

Notes

42 Neuroskeptic, "'Is Your Brain Really Necessary?,' Revisited," *Discover*, July 26, 2015, www.discovermagazine.com/the-sciences/is-your-brain-really-necessary-revisited. Accessed May 7, 2024.

43 Stephen G. Post, *Dignity for Deeply Forgetful People: How Caregivers Can Meet the Challenges of Alzheimer's Disease* (Baltimore: Johns Hopkins University Press, 2022). Kindle edition, location 59–60.

Chapter 4: When Two Minds Must Share Body Parts

1 Denise Ryan, "Conjoined Twins' Family Dreams of a Better Life (with Video)," *Vancouver Sun*, October 25, 2012, https://vancouversun.com/news/metro/conjoined-twins-family-dreams-of-a-better-life-with-video. Accessed May 7, 2024; J. W. Squair, "Craniopagus: Overview and the Implications of Sharing a Brain," *University of British Columbia's Undergraduate Journal of Psychology* 1 (2012), https://ojs.library.ubc.ca/index.php/ubcujp/article/view/2521. The term "thalamic bridge" may have originated with the twins' pediatric neurosurgeon, Douglas Cochrane.

2 Judith Pyke, director, *Inseparable: Ten Years Joined at the Head*, Canadian Broadcasting Corporation, 2017, https://www.youtube.com/watch?v=iKGMeJt7hec&ab_channel=InformOverload; Inform Overload, "7 Yr Old Twins Share the Same Brain & Can See Through Each Others Eye's," YouTube.

3 Joseph Frankel, "Can Children Born Attached at the Head Share a Mind?," *Newsweek*, December 15, 2017, www.newsweek.com/can-conjoined-twins-share-brain-sense-self-748933. Accessed May 7, 2024.

4 J. Savulescu and I. Persson, "Conjoined Twins: Philosophical Problems and Ethical Challenges," *Journal of Medicine and Philosophy* 41, no. 1 (February 2016): 41–55. doi: 10.1093/jmp/jhv037. Epub 2015 Dec 14. PMID: 26671962; PMCID: PMC4882632.

5 Stephen Becker, "Think: How Conjoined Twins Share Their Senses of Sight and Sound," KERA News, July 21, 2014, www.keranews.org/health-science-tech/2014-07-21/think-how-conjoined-twins-share-their-senses-of-sight-and-sound. Accessed May 7, 2024.

6 Denise Ryan, "Through Her Sister's Eyes: Conjoined B.C. Twins Were Extraordinary from the Beginning," *Ottawa Citizen*, October 24, 2024, https://ottawacitizen.com/news/canada/through-her-sisters-eyes-conjoined-bc-twins-tatiana-and-krista-were-extraordinary-from-the-beginning. Accessed March 27, 2024; Becker, "Think: How Conjoined Twins."

7 Pyke, *Inseparable*.

8 Robert Pasnau, "Thomas Aquinas," *Stanford Encyclopedia of Philosophy* (Spring 2023 Edition), ed. Edward N. Zalta and Uri Nodelman, https://plato.stanford.edu/archives/spr2023/entries/aquinas/. Accessed May 7, 2024.

9 Ken MacQueen, "Tatiana and Krista on the Move," *Maclean's*, February 8, 2010, https://macleans.ca/culture/tatiana-and-krista-on-the-move/. Accessed May 7, 2024.

10 Carnevale et al., "Importance of Angiographic Study in Preoperative Planning of Conjoined Twins: Case Report," *Clinics* (São Paulo) 61, no. 2 (April 2006): 167–70. doi: 10.1590/s1807-59322006000200013. Epub 2006 Apr 25. PMID: 16680335.

11 Carnevale et al., "Importance of Angiographic Study."

12 "Conjoined Twins, Epidemiology," Medscape, August 30, 2022, https://emedicine.medscape.com/article/934680-overview?form=fpf. Accessed May 7, 2024.

13 Rachel Paula Abrahamson, "We Are 22-Year-Old Conjoined Twin Sisters: 7 Things We

Notes

Want You to Know," *Today*, April 24, 2023, www.today.com/parents/family/conjoined-twin-sisters-lupita-carmen-andrade-rcna81135. Accessed May 7, 2024.

14. Vanessa LoBue, "What Twins Can Teach Us About Genetic and Environment Influences," *Psychology Today*, March 16, 2021, www.psychologytoday.com/us/blog/the-baby-scientist/202103/what-twins-can-teach-us-about-genetic-and-environment-influences. Accessed May 7, 2024.
15. Kimmy Fitch, "Hensel Twins: Inside the Lives of Abby and Brittany Hensel," *Ask*, August 8, 2022, www.ask.com/lifestyle/hensel-twins-inside-the-lives-of-abby-and-brittany-hensel. Accessed May 7, 2024.
16. Joe Harker, "Conjoined Twins Opened Up on What Their Life Was Like After Becoming Teachers," Unilad, January 23, 2023, www.unilad.com/community/conjoined-twins-teachers-abby-brittany-hensel-618076-20231103. Accessed May 7, 2024.
17. "5th Grade Staff," Sunnyside Elementary, https://sunnyside.mvpschools.org/directory. Accessed May 7, 2024.
18. Lucy Wallace, "Living a Conjoined Life," BBC News, April 25, 2013, www.bbc.com/news/magazine-22181528. Accessed May 7, 2024.
19. Fitch, "Hensel Twins."
20. Fitch, "Hensel Twins."
21. "George Schappell Biography," IMDB, www.imdb.com/name/nm1581627/bio/. Accessed May 7, 2024.
22. Dulcie Pearce, "Joined at the Head: Conjoined Twins George and Lori Schappell: 'We Have Normal Lives,'" *The Sun*, September 13, 2011, www.thesun.co.uk/news/776690/we-have-normal-lives/. Accessed May 7, 2024.
23. Bob Brown and Susie Whitley, "Conjoined Twins, Together Forever," ABC News, September 6, 2006, https://abcnews.go.com/2020/conjoined-twins-forever/story?id=2401005. Accessed May 7, 2024.
24. "Our Lives as Conjoined Twins," BBC News, May 2, 2004, http://news.bbc.co.uk/2/hi/health/3674453.stm. Accessed May 7, 2024.

Chapter 5: The Human Mind Beyond Death

1. Janice Miner Holden, Bruce Greyson, and Debbie James, eds., *The Handbook of Near-Death Experiences: Thirty Years of Investigation* (Praeger, 2009). Kindle edition, location 2716.
2. M. B. Sabom, *Light & Death: One Doctor's Fascinating Account of Near-Death Experiences* (Grand Rapids, MI: Zondervan, 1998), 41.
3. Sabom, *Light & Death*, 42.
4. Sabom, *Light & Death*, 43–44.
5. Sabom, *Light & Death*, 45.
6. Sabom, *Light & Death*, 45.
7. Sabom, *Light & Death*, 45–46.
8. Denyse O'Leary, "Will Studio's New 'After Death' Be a Hit Like 'Sound of Freedom'?," *Mind Matters News*, October 8, 2023, https://mindmatters.ai/2023/10/will-studios-new-after-death-be-a-hit-like-sound-of-freedom/. Accessed May 7, 2024. The neurosurgeon is identified as Dr. Karl Greene.
9. Sabom, *Light & Death*, 50.
10. *The Day I Died: The Mind, the Brain, and Near-Death Experiences*, video recording,

produced by Kate Broome, BBC, 2002, www.documentarytube.com/videos/the-day-i-died/. Accessed May 7, 2024.

11 Mario Beauregard, "Neuroimaging and Spiritual Practice," in *The Oxford Handbook of Psychology and Spirituality*, ed. Lisa J. Miller (Oxford University Press, 2012), 509–10; Keith Augustine, "Does Paranormal Perception Occur in Near-Death Experiences?," *Journal of Near-Death Studies* 25, no. 4 (2007): 203–36; Robert Todd Carroll, "The Skeptic's Dictionary," www.skepdic.com. Accessed June 22, 2015; Evan Thompson, "Waking, Dreaming, Being: Self and Consciousness," in *Neuroscience, Meditation, and Philosophy* (New York: Columbia University Press, 2014), 371.

12 Holden, Greyson, and James, eds., *Handbook of Near-Death Experiences*. Kindle edition, location 580, table 2.3.

13 Bruce Greyson, "The Incidence of Near-Death Experiences," *Medicine and Psychiatry* 1 (December 1998): 92–99, www.newdualism.org/nde-papers/Greyson/Greyson-_1998-1-92-99.pdf.

14 Kenneth Ring, *Life at Death: A Scientific Investigation of the Near-Death Experience* (Coward McCann, 1980). Kindle edition, location 3880.

15 Ring, *Life at Death*, location 428.

16 Marilyn A. Mendoza, "Aftereffects of the Near Death Experience," *Psychology Today*, March 12, 2018, www.psychologytoday.com/us/blog/understanding-grief/201803/aftereffects-the-near-death-experience. Accessed May 7, 2024.

17 S. Khanna and B. Greyson, "Near-Death Experiences and Spiritual Well-Being," *Journal of Religion and Health* 53, no. 6 (December 2014): 1605–15. doi:10.1007/s10943-013-9723-0. PMID: 23625172.

18 Steve Taylor, "Near-Death Experiences and DMT," *Psychology Today*, October 12, 2018, www.psychologytoday.com/us/blog/out-the-darkness/201810/near-death-experiences-and-dmt. Accessed May 7, 2024.

19 Ring, *Life at Death*, location 3674.

20 D. Wilde and C. D. Murray, "Interpreting the Anomalous: Finding Meaning in Out-of-Body and Near-Death Experiences," *Qualitative Research in Psychology* 7, no. 1 (2010): 57–72. doi.org/10.1080/14780880903304550.

21 N. E. Bush and B. Greyson, "Distressing Near-Death Experiences: The Basics," *Missouri Medicine* 111, no. 6 (November–December 2014): 486–90. PMID: 25665233; PMCID: PMC6173534.

22 Bruce Greyson, *After: A Doctor Explores What Near-Death Experiences Reveal About Life and Beyond* (New York: St. Martin's, 2021). Kindle edition, location 194.

23 Greyson, *After*, location 204.

24 Gregory Shushan, "Near-Death Experiences Have Long Inspired Afterlife Beliefs," *Psyche*, July 12, 2021, https://psyche.co/ideas/near-death-experiences-have-long-inspired-afterlife-beliefs. Accessed May 7, 2024.

25 Shushan, "Near-Death Experiences."

26 See, for example, Carol Zaleski, *Otherworld Journeys: Accounts of Near-Death Experience in Medieval and Modern Times* (Oxford University Press, 1988).

27 Howard Storm, *My Descent into Death: A Second Chance at Life* (Harmony, 2005).

28 Bush and Greyson, "Distressing Near-Death Experiences."

29 A. J. Ayer, "What I Saw When I Was Dead," *Sunday Telegraph*, August 28, 1988,

www.philosopher.eu/others-writings/a-j-ayer-what-i-saw-when-i-was-dead/. Accessed May 7, 2024.
30. Greyson, *After*, location 13.
31. See "Greyson NDE Scale," International Association for Near-Death Studies, June 11, 2022, www.iands.org/research/nde-research/important-research-articles/698-greyson-nde-scale.html. Accessed May 7, 2024.
32. Greyson, *After*, location 65–68.
33. Michael Egnor, "Neuroscience Can't Dismiss Near Death Experiences," *Mind Matters News*, May 28, 2020, https://mindmatters.ai/2020/05/neuroscience-cant-dismiss-near-death-experiences/. Accessed May 7, 2024.
34. Holden, Greyson, and James, eds., *Handbook of Near-Death Experiences*, chapter 9.
35. J. M. Holden, "Veridical Perception in Near-Death Experiences," in *Handbook of Near-Death Experiences*, ed. Holden, Greyson, and James, 185–211.
36. Bruce Greyson, "Seeing Dead People Not Known to Have Died: 'Peak in Darien' Experiences," *Anthropology and Humanism* 35 (2010): 159–71. 10.1111/j.1548-1409.2010.01064.x. https://med.virginia.edu/perceptual-studies/wp-content/uploads/sites/360/2017/01/OTH23_Peak-in-Darien-A-H.pdf. Accessed May 7, 2024.
37. Greyson, *After*, location 136.
38. Talia Wise, "New Angel Studios Film 'After Death' Scores 4th Place at Box Office," *CBN*, October 20, 2023, https://cbn.com/news/entertainment/new-angel-studios-film-after-death-scores-4th-place-box-office. Accessed May 7, 2024.
39. See Denyse O'Leary, "Yes, the Film on Near-Death Experiences Is Another 'Hated Hit,'" *Mind Matters News*, November 10, 2023, https://mindmatters.ai/2023/11/yes-the-film-on-near-death-experiences-is-another-hated-hit/. Accessed May 7, 2024.

Chapter 6: The Skeptics' Turn at the Mic

1. Carl Sagan, "In the Valley of the Shadow," *Parade*, March 10, 1996, cited in LibQuotes, https://libquotes.com/carl-sagan/quote/lbr4p3p. Accessed April 1, 2024.
2. Carl Sagan, *Broca's Brain: Reflections on the Romance of Science* (Random House, 1979), 304.
3. Susan Blackmore, "Birth and the OBE: An Unhelpful Analogy," *Journal of the American Society for Psychical Research* 77 (1983): 229–38, www.susanblackmore.uk/articles/birth-and-the-obe-an-unhelpful-analogy/. Accessed May 7, 2024.
4. Michael Shermer, *Why People Believe Weird Things: Pseudoscience, Superstition, and Other Confusions of Our Time* (New York: Henry Holt, 1997), 80.
5. "Near-Death Experience," RationalWiki, https://rationalwiki.org/wiki/Near-death_experience. Accessed September 2, 2023.
6. Christof Koch, "What Near-Death Experiences Reveal About the Brain," *Scientific American*, June 2020, www.scientificamerican.com/article/what-near-death-experiences-reveal-about-the-brain/. Accessed May 7, 2024.
7. Jeffrey Long and Paul Perry, *Evidence of the Afterlife: The Science of Near-Death Experiences* (New York: HarperCollins, 2010), 166.
8. H. Abramovitch, "An Israeli Account of a Near-Death Experience: A Case Study of Cultural Dissonance," *Journal of Near-Death Studies* 6 (1988): 175–84. doi.org/10.1007/BF01073366.

Notes

9 J. Long, "Near-Death Experience. Evidence for Their Reality," *Missouri Medicine* 111, no. 5 (September–October 2014): 372–80. PMID: 25438351; PMCID: PMC6172100. https://www.ncbi.nlm.nih.gov/pmc/articles/PMC6172100/.

10 P. van Lommel et al., "Near-Death Experience in Survivors of Cardiac Arrest: A Prospective Study in the Netherlands," *Lancet* 358, no. 9298 (December 15, 2001): 2039–45. doi:10.1016/S0140-6736(01)07100-8. Erratum in *Lancet* 359, no. 9313 (April 6, 2002): 1254. PMID: 11755611.

11 Robert Birchard, "What Can Science Tell Us About Death?," New York Academy of Sciences, September 30, 2019, www.scribd.com/document/587285586/What-Can-Science-Tell-Us-About-Death. Accessed May 7, 2024.

12 Stephanie Pappas, "Can Science 'Prove' There's an Afterlife? Netflix Documentary Says Yes," Live Science, January 17, 2021, www.livescience.com/netflix-surviving-death.html. Accessed May 7, 2024; David Wilde and Craig D. Murray, "Interpreting the Anomalous: Finding Meaning in Out-of-Body and Near-Death Experiences," *Qualitative Research in Psychology* 7, no. 1 (2010): 57–72. doi:10.1080/14780880903304550.

13 "Depersonalization / Derealization Disorder," *Psychology Today*, www.psychologytoday.com/us/conditions/depersonalizationderealization-disorder. Accessed May 7, 2024.

14 B. Greyson, review of *With the Eyes of the Mind: An Empirical Analysis of Out-of-Body States*, by Glen O. Gabbard and Stuart W. Twemlow, *Journal of Near-Death Studies* 6 (1988): 185–98. doi.org/10.1007/BF01073367.

15 H. J. Irwin, "The Near-Death Experience as a Dissociative Phenomenon: An Empirical Assessment," *Journal of Near-Death Studies* 12, no. 2 (1993): 95–103.

16 Michael B. Sabom, *Recollections of Death: A Medical Investigation* (New York: HarperCollins, 1982), 165.

17 Sabom, *Recollections of Death*, 170.

18 Bruce Greyson, "Western Scientific Approaches to Near-Death Experiences," *Humanities* 4, no. 4 (2015): 775–96. doi.org/10.3390/h4040775.

19 Sabom, *Recollections of Death*.

20 P. Sartori, *The Near-Death Experiences of Hospitalized Intensive Care Patients: A Five Year Clinical Study* (Lewiston, NY: Edwin Mellen Press, 2008).

21 P. van Lommel, "Getting Comfortable with Near-Death Experiences: Dutch Prospective Research on Near-Death Experiences During Cardiac Arrest," *Missouri Medicine* 111, no. 2 (2014): 126–31. PMID: 30323518; PMCID: PMC6179502.

22 Michael Sabom, *Light and Death: One Doctor's Fascinating Account of Near-Death Experiences* (Grand Rapids, MI: Zondervan, 1998), chapter 7.

23 Sabom, *Recollections of Death*, 156.

24 C. Peinkhofer et al., "The Evolutionary Origin of Near-Death Experiences: A Systematic Investigation," *Brain Communications* 3, no. 3 (2021). fcab132. doi.org/10.1093/braincomms/fcab132 cab132.

25 Chris Carter, *Science and the Near-Death Experience: How Consciousness Survives Death* (Inner Traditions, 2010). Kindle edition, location 3179.

26 Long and Perry, *Evidence of the Afterlife*, 111.

27 Long, "Near-Death Experience."

28 W. J. Serdahely, "Questions for the 'Dying Brain Hypothesis,'" *Journal of Near-Death*

Notes

     Studies 15 (1996): 51–52; E. W. Kelly, "Near-Death Experiences with Reports of Meeting Deceased People," *Death Studies* 25, no. 3 (2001): 229–49. doi.org/10.1080/07481180125967.
29 Long, "Near-Death Experience."
30 Melvin Morse and Paul Perry, *Closer to the Light: Learning from Children's Near-Death Experiences* (New York: Villard, 1990), 104.
31 J. Bancaud et al., "Anatomical Origin of Déjà Vu and Vivid 'Memories' in Human Temporal Lobe Epilepsy," *Brain* 117, Pt 1 (February 1994): 71–90. doi:10.1093/brain/117.1.71. PMID: 8149215.
32 E. Bonilla, "Experiencias cercanas a la muerte. Revisión [Near-death experiences]," *Investigación clínica* 52, no. 1 (March 2011): 69–99. Spanish. PMID: 21614815.
33 C. Martial et al. "The Near-Death Experience Content (NDE-C) Scale: Development and Psychometric Validation," *Consciousness and Cognition* 86 (November 2020): 103049. doi:10.1016/j.concog.2020.103049. Epub 2020 Nov 20. PMID: 33227590.
34 Long, "Near-Death Experience."
35 S. Parnia et al., "AWARE—AWAreness during REsuscitation—A Prospective Study," *Resuscitation* 85, no. 12 (December 2014): 1799–805. doi:10.1016/j.resuscitation.2014.09.004. Epub 2014 Oct 7. PMID: 25301715.
36 John M. Evans, "Near-Death Experiences," letter to *The Lancet* 359, Issue 9323 (June 15, 2002). doi.org/10.1016/S0140-6736(02)08925-0.
37 Parnia et al., "AWARE."
38 Seamus Coyle, "Death: Can Our Final Moment Be Euphoric?," *The Conversation*, February 6, 2020, https://theconversation.com/death-can-our-final-moment-be-euphoric-129648. Accessed May 7, 2024.
39 Sabom, *Recollections of Death*, 172.
40 J. H. Lindley, S. Bryan, and B. Conley, "Near-Death Experiences in a Pacific Northwest American Population: The Evergreen Study," *Anabiosis: The Journal of Near-Death Studies* 1, no. 2 (1981): 104–24; P. van Lommel et al., "Near-Death Experience in Survivors of Cardiac Arrest: A Prospective Study in the Netherlands," *Lancet* 358, no. 9298 (December 15, 2001): 2039–45. doi:10.1016/S0140-6736(01)07100-8. Erratum in *Lancet* 359, no. 9313 (April 6, 2002): 1254. PMID: 11755611.
41 Bruce Greyson, "Are Near-Death Experiences Just Dreams or Hallucinations?," transcript, *Big Think*, May 29, 2022, https://bigthink.com/series/devils-advocate/near-death-experiences-hallucinations/. Accessed May 8, 2024.
42 Robert Martone, "New Clues Found in Understanding Near-Death Experiences," *Scientific American*, September 10, 2019, www.scientificamerican.com/article/new-clues-found-in-understanding-near-death-experiences/. Accessed May 8, 2024.
43 B. B. Collier, "Ketamine and the Conscious Mind," *Anaesthesia* 27, no. 2 (April 1972): 120–34. doi:10.1111/j.1365-2044.1972.tb08186.x. PMID: 5021517., p. 126.
44 S. J. Blackmore and T. S. Troscianko, "The Physiology of the Tunnel," *Journal of Near-Death Studies* 8, no. 1 (1989): 15–28. doi.org/10.1007/BF01076136.
45 Chris Carter, *Science and the Near-Death Experience: How Consciousness Survives Death* (Inner Traditions, 2010), Kindle edition, location 2542. See also M. Ohkado and B. Greyson, "A Comparative Analysis of Japanese and Western NDEs," *Journal of Near-Death Studies* 32 (2014): 187–98.

Notes

46 Susan J. Blackmore, *Dying to Live: Near-Death Experiences* (Lanham, MD: Prometheus, 1993).
47 Steve Taylor, "Near-Death Experiences and DMT: A Neurological Explanation of NDEs Remains Elusive," *Psychology Today*, October 12, 2018, www.psychologytoday.com/us/blog/out-the-darkness/201810/near-death-experiences-and-dmt. Accessed May 8, 2024.
48 Parnia et al., "AWARE."
49 "Aware Study Initial Results Are Published!," IANDS, originally published April 22, 2015, updated February 16, 2022, www.iands.org/news/news/front-page-news/1060-aware-study-initial-results-are-published.html. Accessed May 8, 2024.
50 "Aware Study Initial Results," IANDS.
51 Grossman School of Medicine and NYU Langone Health, "Lucid Dying: Patients Recall Death Experiences During CPR," PR Newswire, November 6, 2022, www.prnewswire.com/news-releases/lucid-dying-patients-recall-death-experiences-during-cpr-301669519.html. Accessed May 8, 2024.
52 Grossman and NYU Langone, "Lucid Dying."
53 Kelly Burch, "I've Studied More Than 5,000 Near Death Experiences. My Research Has Convinced Me Without a Doubt That There's Life After Death," *Business Insider*, August 28, 2023, www.insider.com/near-death-experiences-research-doctor-life-after-death-afterlife-2023-8. Accessed May 8, 2024.
54 Susan Blackmore, quoted in Gideon Lichfield, "The Science of Near-Death Experiences: Empirically Investigating Brushes with the Afterlife," *Atlantic*, April 2015, www.theatlantic.com/magazine/archive/2015/04/the-science-of-near-death-experiences/386231/. Accessed May 8, 2024; Michael Shermer, quoted in Marilyn A. Mendoza, "Aftereffects of the Near Death Experience: Adapting to an 'Exceptional Experience,'" *Psychology Today*, March 12, 2018, https://bit.ly/2PxF3NM. Accessed May 8, 2024.
55 Greyson, *After*, location 11.
56 Gary Habermas, "Evidential Near-Death Experiences," chapter 18 in *Minding the Brain* (Seattle: Discovery Institute Press, 2023), 338.

Chapter 7: Immortality of the Soul Is a Reasonable Belief

1 B. Vandermeersch and O. Bar-Yosef, "The Paleolithic Burials at Qafzeh Cave, Israel," *Paleobiology* (2019): 256–75. doi:10.4000/paleo.4848.
2 A. Ronen, "The Oldest Burials and Their Significance," in *African Genesis: Perspectives on Hominin Evolution*, ed. S. C. Reynolds and A. Gallagher, Cambridge Studies in Biological and Evolutionary Anthropology (Cambridge University Press, 2012), 554–70. doi:10.1017/CBO9781139096164.032.
3 M. Martinón-Torres et al., "Earliest Known Human Burial in Africa," *Nature* 593 (2021): 95–100. doi.org/10.1038/s41586-021-03457-8.
4 D. Johnson, "Do Souls Exist?," *Think* 12, no. 35 (2013): 61–75. doi:10.1017/S1477175613000195.
5 "Death: Process or Event," *Britannica*, www.britannica.com/science/death/Death-process-or-event. Accessed September 5, 2023.
6 Peter Kreeft, "The Case for Life After Death," n.d., www.peterkreeft.com/topics-more/case-for-life-after-death.htm. Accessed August 14, 2023.
7 Kreeft, "Life After Death."
8 A spike in neural activity has been reported around the time of death, consistent with

Notes

a departure from usual brain function. See G. Xu et al., "Surge of Neurophysiological Coupling and Connectivity of Gamma Oscillations in the Dying Human Brain," *Proceedings of the National Academy of Sciences of the United States of America* 120, no. 19 (May 9, 2023): e2216268120. doi:10.1073/pnas.2216268120. Epub 2023 May 1. PMID: 37126719; PMCID: PMC10175832. Note: It happens in animals too, but, consistent with different mental processes, it would have different effects.

9   Kreeft, "Life After Death."
10 Carol Zaleski, *Otherworld Journeys: Accounts of Near-Death Experience in Medieval and Modern Times* (Oxford University Press, 1987).
11 John Timmer, "In Depression Treatment Trials, Placebo Effect Is Growing Stronger," *Ars Technica*, September 26, 2023, https://arstechnica.com/science/2023/09/in-depression-treatment-trials-placebo-effect-is-growing-stronger/. Accessed May 8, 2024; Y. Xu et al., "Growing Placebo Response in TMS Treatment for Depression: A Meta-analysis of 27-Year Randomized Sham-Controlled Trials," *Nature Mental Health* 1 (2023): 792–809. doi.org/10.1038/s44220-023-00118-9.
12 H. Benson et al., "Body Temperature Changes During the Practice of g Tum-mo Yoga," *Nature* 295, no. 5846 (January 21, 1982): 234–36. doi:10.1038/295234a0. PMID: 7035966; S. W. Lazar et al., "Functional Brain Mapping of the Relaxation Response and Meditation," *Neuroreport* 11, no. 7 (May 15, 2000): 1581–85. PMID: 10841380; M. Kozhevnikov et al., "Neurocognitive and Somatic Components of Temperature Increases During g-Tummo Meditation: Legend and Reality," *PLoS One* 8, no. 3 (2013): e58244. doi:10.1371/journal.pone.0058244. Epub 2013 Mar 29. PMID: 23555572; PMCID: PMC3612090.
13 Karma Lekshe, Tsomo, "Children in Himalayan Monasteries," in Vanessa R. Sasson (ed.), *Little Buddhas: Children and Childhoods in Buddhist Texts and Traditions*, AAR Religion, Culture, and History (New York, 2012; online edn, Oxford Academic, January 24, 2013). doi.org/10.1093/acprof:oso/9780199860265.003.0016. Accessed October 25, 2024.
14 Mario Beauregard and Denyse O'Leary, *The Spiritual Brain* (Harper One, 2007), 275; M. Beauregard and V. Paquette, "Neural Correlates of a Mystical Experience in Carmelite Nuns," *Neuroscience Letters* 405, no. 3 (September 25, 2006): 186–90. doi: 10.1016/j.neulet.2006.06.060. Epub 2006 Jul 26. PMID: 16872743.
15 Andrew Newberg, "How an Intense Spiritual Retreat Might Change Your Brain," *Psyche*, August 11, 2021, https://psyche.co/ideas/how-an-intense-spiritual-retreat-might-change-your-brain. Accessed May 8, 2024; N. A. Wintering et al., "Effect of a One-Week Spiritual Retreat on Brain Functional Connectivity: A Preliminary Study," *Religions* 12, no. 1 (2021): 23. doi.org/10.3390/rel12010023.
16 Andrew Newberg, Eugene d'Aquili, and Vince Rause, *Why God Won't Go Away: Brain Science and the Biology of Belief* (New York: Ballantine, 2001), 111.
17 Herbert Benson and Marg Stark, *Timeless Medicine: The Power and Biology of Belief* (New York: Scribner, 1996), 116–17.
18 Benson and Stark, *Timeless Medicine*, 99–100. The relationship between mental stress and high blood pressure (hypertension) is still uncertain. What's no longer controversial is the idea that stress could, in principle, be a factor. See University of Chicago, "Loneliness Linked to High Blood Pressure in Aging Adults," *ScienceDaily*, March 28, 2006, www.sciencedaily.com/releases/2006/03/060328081644.htm. Accessed May 8,

Notes

2024; L. C. Hawkley et al., "Loneliness Is a Unique Predictor of Age-Related Differences in Systolic Blood Pressure," *Psychology and Aging* 21, no. 1 (March 2006): 152-64. doi:10.1037/0882-7974.21.1.152. PMID: 16594800.

19 Benson and Stark, *Timeless Medicine*, 152.
20 Jeff Levin and Harold G. Koenig, eds., *Faith, Medicine, and Science: A Festschrift in Honor of Dr. David B. Larson* (New York: Haworth, 2005), 19.
21 Levin and Koenig, *Faith, Medicine, and Science*, 85.
22 D. A. Matthews et al., *The Faith Factor: An Annotated Bibliography of Clinical Research on Spiritual Subjects* (National Institute for Healthcare Research, 1993).
23 Sabom, *Light and Death*, 81–82.
24 K. I. Pargament et al., "Religious Struggle as a Predictor of Mortality Among Medically Ill Elderly Patients: A 2-Year Longitudinal Study," *Archives of Internal Medicine* 161, no. 15 (August 13–27, 2001): 1881–85. doi:10.1001/archinte.161.15.1881. PMID: 11493130.
25 Baylor University, "Study Examines Link Between Accountability to God and Psychological Well-Being," *Neuroscience News*, March 2, 2022, https://neurosciencenews.com/religious-accountability-wellbeing-20126/. Accessed May 8, 2024; M. Bradshaw et al., "Perceptions of Accountability to God and Psychological Well-Being Among US Adults," *Journal of Religion and Health* 61, no. 1 (February 2022): 327–52. doi:10.1007/s10943-021-01471-8. Epub 2022 Jan 18. PMID: 35039960. Top of Form.
26 Eric W. Dolan, "New Study Links Intrinsic Religious Motivation to Higher-Level Patterns of Thought," *PsyPost*, May 22, 2021, www.psypost.org/2021/05/new-study-links-intrinsic-religious-motivation-to-higher-level-patterns-of-thought-60857. Accessed May 8, 2024; J. L. Michaels, J. Petrino, and T. Pitre-Zampol, "Individual Differences in Religious Motivation Influence How People Think," *Journal for the Scientific Study of Religion* 60 T. (2021): 64–82. doi.org/10.1111/jssr.12696.
27 H. M. Helm et al., "Does Private Religious Activity Prolong Survival? A Six-Year Follow-Up Study of 3,851 Older Adults," *Journals of Gerontology Series A: Biological Sciences and Medical Sciences* 55, no. 7 (July 2000): M400-5. doi:10.1093/gerona/55.7.m400. PMID: 10898257.
28 Mike Emery, "Power of the Pulpit: Study Suggests Lower Mortality Rates for Black American Men 50+ Who Attend Religious Services," *Medical XPress*, September 23, 2022, https://medicalxpress.com/news/2022-09-power-pulpit-mortality-black-men.html. Accessed May 8, 2024; M. A. Bruce et al., "Religious Service Attendance and Mortality Among Older Black Men," *PLoS One* 17, no. 9 (September 2, 2022): e0273806. doi:10.1371/journal.pone.0273806. PMID: 36054189; PMCID: PMC9439243.
29 Andrew Newberg, "Mind and God: The New Science of Neurotheology," *Big Think*, May 6, 2021, https://bigthink.com/neuropsych/mind-god-new-science-neurotheology/. Accessed May 8, 2024.
30 Sabom, *Light and Death*, 126.
31 See 2 Corinthians 12:1–4. Paul is generally thought to be referring to himself here.
32 Neil Dagnall and Ken Drinkwater, "Are Near-Death Experiences Hallucinations? Experts Explain the Science Behind This Puzzling Phenomenon," *The Conversation*, December 7, 2018, https://theconversation.com/are-near-death-experiences-hallucinations-experts-explain-the-science-behind-this-puzzling-phenomenon-106286. Accessed May 8, 2024.

Notes

33 See, for example, C. J. Kazilek and Kim Cooper, "ASU: Ask a Biologist: Colors Animals See," December 17, 2009, https://askabiologist.asu.edu/colors-animals-see. Accessed November 3, 2023.
34 Habermas, "Evidential Near-Death Experiences," 214.
35 K. Ring and S. Cooper, "Near-Death and Out-of-Body Experiences in the Blind: A Study of Apparent Eyeless Vision," *Journal of Near-Death Studies* 16, no. 2 (1997): 101–47. doi.org/10.1023/A:1025010015662.

Chapter 8: Free Will Is a Real and Intrinsic Part of the Soul

1 Timothy Revell, "Why Free Will Doesn't Exist, According to Robert Sapolsky," *New Scientist*, October 18, 2023, www.newscientist.com/article/2398369-why-free-will-doesnt-exist-according-to-robert-sapolsky/. Accessed May 8, 2024.
2 See Daniel Menaker, "Have It Your Way," *New York Times*, July 13, 2012. www.nytimes.com/2012/07/15/books/review/free-will-by-sam-harris.html. Accessed May 8, 2024.
3 Massimo Pigliucci, "Consciousness, Decision Making, and 'Free' Will," *Medium*, February 7, 2022, https://philosophyasawayoflife.medium.com/consciousness-decision-making-and-free-will-94a0724fc70b. Accessed May 8, 2024.
4 Jerry A. Coyne, "You Don't Have Free Will," *Chronicle of Higher Education*, March 18, 2012, www.chronicle.com/article/you-dont-have-free-will/. Accessed May 8, 2024.
5 Avi Loeb, "How Much Time Does Humanity Have Left?," *Scientific American*, May 12, 2021, www.scientificamerican.com/article/how-much-time-does-humanity-have-left/. Accessed May 8, 2024.
6 David Eagleman, "The Brain on Trial," *Atlantic*, June 7, 2011, www.theatlantic.com/magazine/archive/2011/07/the-brain-on-trial/308520/. Accessed May 8, 2024.
7 Sabine Hossenfelder, "How to Live Without Free Will," *BackRe(Action)*, May 2, 2019, https://backreaction.blogspot.com/2019/05/how-to-live-without-free-will.html. Accessed May 8, 2024.
8 Larry Krauss, "Philosophy v Science: Which Can Answer the Big Questions of Life," *Guardian*, September 9, 2012, www.theguardian.com/science/2012/sep/09/science-philosophy-debate-julian-baggini-lawrence-krauss. Accessed May 8, 2024.
9 Yuval Noah Harari, "The Myth of Freedom," *Guardian*, September 14, 2018, www.theguardian.com/books/2018/sep/14/yuval-noah-harari-the-new-threat-to-liberal-democracy. Accessed May 8, 2024.
10 Mustafa Gatollari, "Yes, David Berkowitz Blamed His Serial Killing Urgings on a Labrador Retriever," *Distractify*, May 5, 2021, www.distractify.com/p/son-of-sam-neighbor-dog. Accessed May 8, 2024.
11 See, for example, Hannah Arendt, "What Is Freedom?," in *Between Past and Future*, reprinted in *Philosophical Notebooks*, August 27, 2017, https://philosophicalnotebooks.wordpress.com/2017/08/27/arendt-what-is-freedom/. Accessed May 3, 2024.
12 P. E. Tressoldi, "Extraordinary Claims Require Extraordinary Evidence: The Case of Non-local Perception, a Classical and Bayesian Review of Evidences," *Frontiers in Psychology* 10, no. 2 (June 2011): 117. doi:10.3389/fpsyg.2011.00117. PMID: 21713069; PMCID: PMC3114207.
13 Quoted in Robert N. Goldman, ed., *Einstein's God: Albert Einstein's Quest as a Scientist*

Notes

and as a Jew to Replace a Forsaken God (Jason Aronson Inc., 1997), cited in LibQuotes, https://libquotes.com/albert-einstein/quote/lbb2g5n. Accessed May 3, 2024.

14 As recounted in C. P. Snow, *The Physicists: A Generation That Changed the World* (Macmillan, 1981), 84.

15 Wilder Penfield, *The Mystery of the Mind: A Critical Study of Consciousness and the Human Brain* (Princeton Legacy Library, 2015), 76–77.

16 Benjamin Libet, *Mind Time: The Temporal Factor in Consciousness* (Harvard University Press, 2005). Libet discusses his free will experiments in chapter 4.

17 A. Schurger et al., "What Is the Readiness Potential?," *Trends in Cognitive Science* 25, no. 7 (July 2021): 558–70. doi:10.1016/j.tics.2021.04.001. Epub 2021 Apr 27. PMID: 33931306; PMCID: PMC8192467.

18 Julian Baggini, "How to Think About Free Will," *Psyche*, May 11, 2022, https://psyche.co/guides/how-to-think-about-free-will-in-a-world-of-cause-and-effect. Accessed May 8, 2024.

19 B. Libet, "Do We Have Free Will?," *Journal of Consciousness Studies* 6, nos. 8–9 (1999): 47–57; B. Libet, "The Neural Time—Factor in Perception, Volition and Free Will," in *Neurophysiology of Consciousness. Contemporary Neuroscientists* (Boston: Birkhäuser, 1993). doi.org/10.1007/978-1-4612-0355-1_22.

20 Benjamin Libet, "Do We Have Free Will?," in *Conscious Will and Responsibility: A Tribute to Benjamin Libet*, ed. Walter Sinnott-Armstrong and Lynn Nadel (2010; online ed., Oxford Academic, January 1, 2011). doi.org/10.1093/acprof:oso/9780195381641.003.0002. Accessed December 18, 2023.

21 Benjamin Libet, *Mind Time: The Temporal Factor in Consciousness* (Harvard University Press, 2005), 151.

22 Bahar Gholipour, "Famous Argument Against Free Will Debunked," *Atlantic*, September 10, 2019, www.theatlantic.com/health/archive/2019/09/free-will-bereitschaftspotential/597736/. Accessed May 8, 2024; Schurger et al., "What Is the Readiness Potential?"

23 Alessandra Buccella and Tomáš Dominik, "Free Will Is Only an Illusion If You Are, Too," *Scientific American*, January 16, 2023, www.scientificamerican.com/article/free-will-is-only-an-illusion-if-you-are-too/. Accessed May 8, 2024; U. Maoz, G. Yaffe, C. Koch, and L. Mudrik, "Neural Precursors of Decisions That Matter—an ERP Study of Deliberate and Arbitrary Choice," *eLife* 8 (October 23, 2019): e39787. doi:10.7554/eLife.39787. PMID: 31642807; PMCID: PMC6809608.

24 Cristi L. S. Cooper, "Free Will, Free Won't, and What the Libet Experiments Don't Tell Us," chapter 14 in *Minding the Brain: Models of the Mind, Information, and Material Science* (Seattle: Discovery Institute Press, 2023), 271.

25 Victor Madeira, "Judgment in Moscow: Soviet Crimes and Western Complicity," *International Affairs* 96, Issue 1 (January 2020): 253–55. doi.org/10.1093/ia/iiz176.

26 George Ellis, "From Chaos to Free Will," *Aeon*, June 9, 2020, https://aeon.co/essays/heres-why-so-many-physicists-are-wrong-about-free-will. Accessed May 8, 2024.

27 John Horgan, "Free Will Is Real," *Scientific American*, June 3, 2019, https://blogs.scientificamerican.com/cross-check/free-will-is-real/. Accessed May 8, 2024.

28 John Horgan, "Does Quantum Mechanics Rule Out Free Will?," *Scientific American*, March 10, 2022, www.scientificamerican.com/article/does-quantum-mechanics-rule-out-free-will/. Accessed May 8, 2024.

Notes

29 John Horgan, "Free Will and ChatGPT-Me," *Cross-Check*, November 16, 2023, https://johnhorgan.org/cross-check/free-will-and-chatgpt-me. Accessed May 8, 2024.
30 Noam Chomsky, "Noam Chomsky's Philosophy: Free Will 1," YouTube, September 29, 2014, www.youtube.com/watch?v=J3fhKRJNNTA. Accessed September 9, 2023.
31 George Musser, "Yes, Determinists, There Is Free Will," *Nautilus*, May 13, 2019, https://nautil.us/yes-determinists-there-is-free-will-237396/. Accessed May 8, 2024.
32 D. Bowen, "Daniel Dennett on Free Will," in: *Encyclopedia of Evolutionary Psychological Science*, ed. T. Shackelford and V. Weekes-Shackelford (Springer Cham, 2018). doi.org/10.1007/978-3-319-16999-6_2168-1.
33 Tammie Sommers, "Can an Atheist Believe in Free Will?," *Psychology Today*, January 22, 2009, www.psychologytoday.com/us/blog/experiments-in-philosophy/200901/can-atheist-believe-in-free-will. Accessed May 8, 2024.
34 Steven Pinker, X [Twitter], November 10, 2023, https://twitter.com/sapinker/status/1723038526508310785. Accessed May 8, 2024.
35 Kevin J. Mitchell, "How Life Evolved the Power to Choose," Princeton University Press, October 27, 2023, https://press.princeton.edu/ideas/how-life-evolved-the-power-to-choose. Accessed May 8, 2024.
36 Thomas Aquinas, "Whether There Is in Us a Natural Law?," Summa Theologica, https://ccel.org/ccel/aquinas/summa.FS_Q91_A2.html. Accessed May 8, 2024.

Chapter 9: Models of the Mind—Which One Fits Best?

1 Mariana Lenharo, "Decades-Long Bet on Consciousness Ends—and It's Philosopher 1, Neuroscientist 0," *Nature*, June 24, 2023, www.nature.com/articles/d41586-023-02120-8.
2 Michael D. Lemonick, "Glimpses of the Mind," *Time*, July 17, 1995, https://content.time.com/time/subscriber/article/0,33009,983176,00.html. Accessed May 8, 2024.
3 Steven Pinker, "The Mystery of Consciousness," *Time*, January 19, 2007, www.psy.vanderbilt.edu/courses/hon182/Mystery_of_consciousness_Time_Mag_2007_Pinker.pdf. Accessed May 8, 2024.
4 Massimo Pigliucci, "Consciousness Is Real," *Aeon*, December 16, 2019, https://aeon.co/essays/consciousness-is-neither-a-spooky-mystery-nor-an-illusory-belief. Accessed May 8, 2024.
5 Noam Chomsky, *Syntactic Structures* (Berlin: Mouton, 1957), 15.
6 Steven Schneider, "Identity Theory," *Internet Encyclopedia of Philosophy*. Accessed December 18, 2023.
7 Georges Rey, "The Computational-Representational Theory of Thought (CRTT)," *Britannica*, www.britannica.com/topic/philosophy-of-mind/The-computational-representational-theory-of-thought-CRTT. Accessed December 18, 2023.
8 J. R. Searle, "Minds, Brains, and Programs," *Behavioral and Brain Sciences* 3, no. 3 (1980): 417–24. doi.org/10.1017/S0140525X00005756.
9 "Chinese Room Argument, *Britannica*, www.britannica.com/topic/Chinese-room-argument. Accessed May 8, 2024.
10 "Functionalism," *Britannica*, www.britannica.com/topic/functionalism-philosophy-of-mind. Accessed May 8, 2024.
11 William Ramsey, "Eliminative Materialism," *Stanford Encyclopedia of Philosophy* (Spring 2022 Edition), ed. Edward N. Zalta, https://plato.stanford.edu/archives/spr2022/entries/materialism-eliminative/. Accessed May 8, 2024.

Notes

12   John Bickle, Peter Mandik, and Anthony Landreth, "The Philosophy of Neuroscience," *Stanford Encyclopedia of Philosophy* (Fall 2019 Edition), ed. Edward N. Zalta, https://plato.stanford.edu/archives/fall2019/entries/neuroscience. Accessed May 8, 2024.

13   Patricia Churchland, "The Big Questions: Do We Have Free Will?," NewScientist.com news service, November 18, 2006, https://www.newscientist.com/article/mg19225780-070-the-big-questions-do-we-have-free-will/. Accessed May 8, 2024.

14   Joe Gough, "The Mind Does Not Exist," *Aeon*, August 30, 2021, https://aeon.co/essays/why-theres-no-such-thing-as-the-mind-and-nothing-is-mental. Accessed May 8, 2024.

15   Michael W. Taft, "What Is the Self? An Interview with Thomas Metzinger," *Deconstructing Yourself*, September 10, 2017, https://deconstructingyourself.com/what-is-the-self-metzinger.html. Accessed May 8, 2024.

16   Peter L. Halligan and David A. Oakley, "Is It Time to Give Up on Consciousness as 'The Ghost in the Machine'?," *The Conversation*, June 6, 2021, https://theconversation.com/is-it-time-to-give-up-on-consciousness-as-the-ghost-in-the-machine-160688. Accessed May 8, 2024.

17   Walter Velt, "Disenchanted Naturalism, Part 3," *Psychology Today*, March 20, 2020 (interview with Alex Rosenberg), www.psychologytoday.com/intl/blog/science-and-philosophy/202003/disenchanted-naturalism. Accessed May 8, 2024.

18   Donald Hoffman, "Is Reality Real? How Evolution Blinds Us to the Truth About the World," *New Scientist*, July 31, 2019, www.newscientist.com/article/mg24332410-300-is-reality-real-how-evolution-blinds-us-to-the-truth-about-the-world/. Accessed May 8, 2024.

19   Edward Feser, "Mad Dogs and Eliminativists," *Edward Feser* (blog), August 21, 2013, https://edwardfeser.blogspot.com/2013/08/mad-dogs-and-eliminativists.html. Accessed May 8, 2024.

20   P. Carruthers, "The Illusion of Conscious Thought," *Journal of Consciousness Studies* 24, no. 9–10 (2017): 228–52.

21   Edward Feser, "Some Brief Arguments for Dualism, Part II," *Edward Feser*, September 26, 2008, https://edwardfeser.blogspot.com/2008/09/some-brief-arguments-for-dualism-part.html. Accessed May 8, 2024.

22   Roger Scruton, "Brain Drain: Neuroscience Wants to Be the Answer to Everything. It Isn't," *Spectator*, March 17, 2012, www.spectator.co.uk/article/brain-drain/. Accessed May 8, 2024.

23   M. R. Bennett and P. M. S. Hacker, *Philosophical Foundations of Neuroscience* (Malden, MA: Blackwell, 2003), 275/1079 e-book.

24   "Consciousness," *Stanford Encyclopedia of Philosophy*, https://plato.stanford.edu/entries/consciousness/, updated Jan 14, 2014. Accessed May 8, 2024.

25   Philip Goff, William Seager, and Sean Allen-Hermanson, "Panpsychism," *Stanford Encyclopedia of Philosophy* (Summer 2022 Edition), ed. Edward N. Zalta, https://plato.stanford.edu/archives/sum2022/entries/panpsychism/.

26   The categories chosen are some of those used in "2.1 Creature Consciousness," "Consciousness," *Stanford Encyclopedia of Philosophy*, https://plato.stanford.edu/entries/consciousness/#CreCon.

27   M. Lenharo, "Decades-Long Bet on Consciousness Ends—and It's Philosopher 1,

Notes

Neuroscientist 0," *Nature* 619, no. 7968 (July 2023): 14–15. doi:10.1038/d41586-023-02120-8. PMID: 37353639.

28 Francis Fallon, "Integrated Information Theory of Consciousness," *Internet Encyclopedia of Philosophy*, https://iep.utm.edu/integrated-information-theory-of-consciousness/. Accessed July 14, 2024.

29 IIT-Concerned, Stephen M. Fleming, Chris Frith, Mel Goodale, Hakwan Lau, Joseph E. LeDoux, Alan L. F. Lee, et al. "The Integrated Information Theory of Consciousness as Pseudoscience," PsyArXiv, September 16, 2023. doi:10.31234/osf.io/zsr78.

30 Erik Hoel, "No Current Theory of Consciousness is Scientific," IAI News, September 25, 2023, https://iai.tv/articles/no-theory-of-consciousness-is-scientific-auid-2610?_auid=2020. Accessed October 19, 2023.

31 John Horgan, "The Brouhaha over Consciousness and 'Pseudoscience,'" *Cross-Check*, September 23, 2023, https://johnhorgan.org/cross-check/the-brouhaha-over-consciousness-and-pseudoscience. Accessed October 19 , 2023.

Chapter 10: The Human Mind Has No History

1 C. P. Rajendran, "Decoding the Mystery of Consciousness—From Hominids to Humans," *Science: The Wire*, April 11, 2020, https://science.thewire.in/the-sciences/evolution-human-intelligence-consciousness-social-order-brain-size/. Accessed May 9, 2024.

2 Stephen Jay Gould, *Ever Since Darwin: Reflections in Natural History* (Norton, 1977), 50.

3 Ralph Lewis, "Is Consciousness an Illusion?, Part 5," *Psychology Today*, October 22, 2020, www.psychologytoday.com/ca/blog/finding-purpose/202010/is-consciousness-illusion. Accessed May 9, 2024.

4 "While the genetic difference between individual humans today is minuscule—about 0.1%, on average—study of the same aspects of the chimpanzee genome indicates a difference of about 1.2%." From "What Does It Mean to Be Human?: Genetics," Smithsonian Museum of Natural History, https://humanorigins.si.edu/evidence/genetics. Accessed December 19, 2023.

5 L. Marino, "A Comparison of Encephalization Between Odontocete Cetaceans and Anthropoid Primates," *Brain, Behavior and Evolution* 51, no. 4 (1998): 230–38. doi: 10.1159/000006540. Erratum in *Brain, Behavior and Evolution* 52, no. 1 (1998): 22. PMID: 9553695; S. Benito-Kwiecinski, S. L. Giandomenico, M. Sutcliffe, E. S. Riis, P. Freire-Pritchett, I. Kelava, S. Wunderlich, U. Martin, G. A. Wray, K. McDole, and M. A. Lancaster, "An Early Cell Shape Transition Drives Evolutionary Expansion of the Human Forebrain," *Cell* 184, no. 8 (April 15, 2021): 2084–102.e19. doi:10.1016/j.cell.2021.02.050. Epub 2021 Mar 24. PMID: 33765444; PMCID: PMC8054913.

6 S. Herculano-Houzel et al., "The Elephant Brain in Numbers," *Frontiers in Neuroanatomy* 8 (June 12, 2014): 46. doi:10.3389/fnana.2014.00046. PMID: 24971054; PMCID: PMC4053853.

7 Rachel Nuwer, "Young Ravens Rival Adult Chimps in a Big Test of General Intelligence," *Scientific American*, December 10, 2020, www.scientificamerican.com/article/young-ravens-rival-adult-chimps-in-a-big-test-of-general-intelligence/. Accessed May 9, 2024.

Notes

8   Finkel, "How the Octopus Got Its Smarts," *Cosmos*, September 16, 2018, https://cosmosmagazine.com/nature/how-the-octopus-got-its-smarts/. Accessed May 9, 2024.
9   Finkel, "How the Octopus Got Its Smarts."
10 Ed Yong, "For Smart Animals, Octopuses Are Very Weird," *Atlantic*, July 2, 2019, www.theatlantic.com/science/archive/2019/07/why-did-octopuses-become-smart/593155/. Accessed May 9, 2024.
11 Colin Barras, "Cats Know Their Names—Whether They Care Is Another Matter," *Nature*, April 4, 2019, www.nature.com/articles/d41586-019-01067-z. Accessed May 9, 2024; A. Saito et al., "Domestic Cats (*Felis catus*) Discriminate Their Names from Other Words," *Scientific Reports* 9, no. 1 (April 4, 2019): 5394. doi:10.1038/s41598-019-40616-4. Erratum in *Scientific Reports* 9, no. 1 (September 10, 2019): 13265. PMID: 30948740; PMCID: PMC6449508.
12 Ashley P. Taylor, "Why Do Parrots Talk?," *Audubon*, August 6, 2015, www.audubon.org/news/why-do-parrots-talk. Accessed May 9, 2024.
13 Sophie Fessl, "Can These Fish Do Math?," *The Scientist*, March 31, 2022, www.the-scientist.com/news-opinion/can-these-fish-do-math-69861. Accessed May 9, 2024; V. Schluessel et al., "Cichlids and Stingrays Can Add and Subtract 'One' in the Number Space from One to Five," *Scientific Reports* 12, no. 1 (March 31, 2022): 3894. doi:10.1038/s41598-022-07552-2. PMID: 35361791; PMCID: PMC8971382.
14 Max Planck Society, "The Algebra of Neurons: Study Deciphers How a Single Nerve Cell Can Multiply," Phys.org, February 23, 2022, https://phys.org/news/2022-02-algebra-neurons-deciphers-nerve-cell.html. Accessed May 9, 2024; L. N. Groschner et al., "A Biophysical Account of Multiplication by a Single Neuron," *Nature* 603, no. 7899 (March 2022): 119–23. doi:10.1038/s41586-022-04428-3. Epub 2022 Feb 23. PMID: 35197635; PMCID: PMC8891015.
15 Scarlett Howard et al., "Honeybees Join Humans as the Only Known Animals That Can Tell the Difference Between Odd and Even Numbers," Phys.org, April 29, 2022, https://phys.org/news/2022-04-honeybees-humans-animals-difference-odd.html. Accessed May 9, 2024; Scarlett R. Howard et al., "Numerosity Categorization by Parity in an Insect and Simple Neural Network," *Frontiers in Ecology and Evolution* 10 (2022). doi.org/10.3389/fevo.2022.805385.
16 Brian Butterworth, "A Basic Sense of Numbers Is Shared by Countless Creatures," *Psyche*, October 12, 2022, https://psyche.co/ideas/a-basic-sense-of-numbers-is-shared-by-countless-creatures. Accessed May 9, 2024.
17 "Are Our Neurons Really Wired for Numbers?," *Mind Matters News*, October 11, 2021, https://mindmatters.ai/2021/10/are-our-neurons-really-wired-for-numbers/. Accessed May 9, 2024.
18 Colin Barras, "How Did Neanderthals and Other Ancient Humans Learn to Count?," *Nature*, June 2, 2021, www.nature.com/articles/d41586-021-01429-6. Accessed May 9, 2024.
19 Margaret Wertheim, "Radical Dimensions," *Aeon*, January 10, 2018, https://aeon.co/essays/how-many-dimensions-are-there-and-what-do-they-do-to-reality. Accessed May 9, 2024.
20 Robert J. Marks, "Why Infinity Does Not Exist in Reality," *Mind Matters News*, July 1, 2022, https://mindmatters.ai/2022/07/1-why-infinity-does-not-exist-in-reality/. Accessed May 9, 2024.

Notes

21 Jane C. Hu, "What Do Talking Apes Really Tell Us?," *Slate*, August 20, 2014, https://slate.com/technology/2014/08/koko-kanzi-and-ape-language-research-criticism-of-working-conditions-and-animal-care.html. Accessed May 9, 2024.

22 Derrick Bryson Taylor, "Border Collie Trained to Recognize 1,022 Nouns Dies," *New York Times*, July 27, 2019, www.nytimes.com/2019/07/27/science/chaser-border-collie-dies.html. Accessed May 9, 2024.

23 University of Portsmouth, "Environment, Not Evolution, Might Underlie Some Human-Ape Differences," *ScienceDaily*, July 15, 2019, www.sciencedaily.com/releases/2019/07/190715094847.htm. Accessed May 9, 2024; D. A. Leavens, K. A. Bard, and W. D. Hopkins, "The Mismeasure of Ape Social Cognition," *Animal Cognition* 22, no. 4 (July 2019): 487–504. doi:10.1007/s10071-017-1119-1. Epub 2017 Aug 4. PMID: 28779278; PMCID: PMC6647540.

24 Rachel Nuwer, "To Communicate with Apes, We Must Do It on Their Terms," *Nova*, April 25, 2018, www.pbs.org/wgbh/nova/article/to-communicate-with-apes-we-must-do-it-on-their-terms/. Accessed May 9, 2024.

25 David P. Barash, "Even Worms Feel Pain," *Nautilus*, March 2, 2022, https://nautil.us/even-worms-feel-pain-238436/. Accessed May 9, 2024.

26 Natalie Mesa, "Do Invertebrates Have Emotions?," *The Scientist*, May 26, 2022, www.the-scientist.com/news-opinion/do-invertebrates-have-emotions-70067. Accessed May 9, 2024.

27 Jordana Cepelewicz, "Animals Count and Use Zero. How Far Does Their Number Sense Go?," *Quanta*, August 9, 2021, www.quantamagazine.org/animals-can-count-and-use-zero-how-far-does-their-number-sense-go-20210809/. Accessed May 9, 2024; M. E. Kirschhock, H. M. Ditz, and A. Nieder, "Behavioral and Neuronal Representation of Numerosity Zero in the Crow," *Journal of Neuroscience* 41, no. 22 (June 2, 2021): 4889–96. doi:10.1523/JNEUROSCI.0090-21.2021. Epub 2021 Apr 19. PMID: 33875573; PMCID: PMC8260164.

28 Denise Flaim, "Akita History: Hachikō & the Revival of the Devoted Japanese Breed," American Kennel Club, February 18, 2021, www.akc.org/expert-advice/dog-breeds/akita-history-hachiko-japanese-breed/. Accessed May 9, 2024.

29 Neil Thomas, "How Darwin and Wallace Split over the Human Mind," *Evolution News*, June 13, 2022, https://evolutionnews.org/2022/06/how-darwin-and-wallace-split-over-the-human-mind/. Accessed May 9, 2024.

30 Brandon Keim, "Poop-Throwing Chimps Provide Hints of Human Origins," *Wired Science*, November 29, 2011, www.wired.com/2011/11/chimp-throwing/. Accessed May 9, 2024.

31 Nora Schultz, "Baby Apes' Arm Waving Hints at Origins of Language," *New Scientist*, November 10, 2011, www.newscientist.com/article/dn21152-baby-apes-arm-waving-hints-at-origins-of-language/. Accessed May 9, 2024.

32 The Editors, "In Praise of... Neanderthal Man," *Guardian*, January 13, 2010, www.theguardian.com/commentisfree/2010/jan/13/in-praise-of-neanderthal-man. Accessed May 9, 2024.

33 Lin Edwards, "Evidence Neanderthals Used Feathers for Decoration," Phys.org, February 23, 2011, https://phys.org/news/2011-02-evidence-neanderthals-feathers.html. Accessed May 9, 2024.

34 Bob Yirka, "Neanderthal Home Made of Mammoth Bones Discovered in Ukraine,"

Notes

Phys.org, December 19, 2011, https://phys.org/news/2011-12-neanderthal-home-mammoth-bones-ukraine.html. Accessed May 9, 2024.

35 University of Exeter, "New Evidence Debunks 'Stupid' Neanderthal Myth," *ScienceDaily*, August 26, 2008, www.sciencedaily.com/releases/2008/08/080825203924.htm. Accessed May 9, 2024.

36 Tom Worden, "'The Oldest Work of Art Ever': 42,000-Year-Old Paintings of Seals Found in Spanish Cave," *Daily Mail*, February 7, 2012, www.dailymail.co.uk/sciencetech/article-2097869/The-oldest-work-art-42-000-year-old-paintings-seals-Spanish-cave.html. Accessed May 9, 2024.

37 B. Wood, "Did Early Homo Migrate 'Out of' or 'In to' Africa?," *Proceedings of the National Academy of Sciences of the United States of America* 108, no. 26 (2011): 10375–76. doi.org/10.1073/pnas.1107724108.

38 Bruce Bower, "A Child's 78,000-Year-Old Grave Marks Africa's Oldest Known Human Burial," *Science News*, May 5, 2021, www.sciencenews.org/article/africa-oldest-known-human-burial-child-grave-cave. Accessed May 9, 2024; M. Martinón-Torres et al., "Earliest Known Human Burial in Africa," *Nature* 593, no. 7857 (May 2021): 95–100. doi:10.1038/s41586-021-03457-8. Epub 2021 May 5. PMID: 33953416. *Mtoto* is Swahili for "child."

39 Bruce Bower, "Stone Age Code Red: Scarlet Symbols Emerge in Israeli Cave," *Science News*, October 29, 2003, www.sciencenews.org/article/stone-age-code-red-scarlet-symbols-emerge-israeli-cave. Accessed May 9, 2024.

40 K. Kris Hirst, "Qafzeh Cave, Israel: Evidence for Middle Paleolithic Burials," ThoughtCo, November 17, 2019, www.thoughtco.com/qafzeh-cave-israel-middle-paleolithic-burials-172284. Accessed May 9, 2024.

41 PLOS, "3-D Image of Paleolithic Child's Skull Reveals Trauma, Brain Damage," *ScienceDaily*, July 23, 2014, www.sciencedaily.com/releases/2014/07/140723141714.htm.

42 Bower, "Stone Age Code Red."

43 Phie Jacobs, "Were These Stone Balls Made by Ancient Human Relatives Trying to Perfect the Sphere?," *Science*, September 5, 2023, www.science.org/content/article/were-these-stone-balls-made-ancient-human-relatives-trying-perfect-sphere. Accessed May 9, 2024; A. Muller et al., "The Limestone Spheroids of 'Ubeidiya: Intentional Imposition of Symmetric Geometry by Early Hominins?," *Royal Society Open Science* 10, no. 9 (September 6, 2023): 230671. doi:10.1098/rsos.230671. PMID: 37680494; PMCID: PMC10480702.

44 Charles C. Mann, "The Birth of Religion," *National Geographic*, June 2011, www.nationalgeographic.com/magazine/article/gobeki-tepe. Accessed May 9, 2024.

45 Mann, "The Birth of Religion."

46 Michael Egnor, "Aristotle on the Immateriality of Intellect and Will," *Evolution News*, January 26, 2015, https://evolutionnews.org/2015/01/aristotle_on_th/. Accessed May 9, 2024.

Chapter 11: What Does It All Mean? Neuroscience Meets Philosophy

1 E. Feser, *Aquinas: A Beginner's Guide* (Simon & Schuster, 2009), Kindle edition, location 180.

2 Feser, *Aquinas*, location 202.

3 Peter L. Halligan and David A. Oakley, "Is It Time to Give Up on Consciousness

Notes

as 'The Ghost in the Machine'?," *The Conversation*, June 6, 2021, https://theconversation.com/is-it-time-to-give-up-on-consciousness-as-the-ghost-in-the-machine-160688. Accessed May 9, 2024.

4 Gert-Jan Lokhorst, "Descartes and the Pineal Gland," *Stanford Encyclopedia of Philosophy* (Winter 2021 Edition), ed. Edward N. Zalta, https://plato.stanford.edu/archives/win2021/entries/pineal-gland/. Accessed May 9, 2024.

5 Noelle Trent, "Frederick Douglass," *Britannica*, May 8, 2024, www.britannica.com/biography/Frederick-Douglass. Accessed May 9, 2024.

6 Feser, *Aquinas*, location 2,541.

Chapter 12: And This All Men Call God

1 Jonathan Haidt, "Have We Evolved to Be Religious?," *Time*, March 27, 2012. https://ideas.time.com/2012/03/27/have-we-evolved-to-be-religious/. Accessed May 9, 2024.

2 I. Pyysiäinen and M. Hauser, "The Origins of Religion: Evolved Adaptation or By-Product?," *Trends in Cognitive Sciences* 14, no. 3 (March 2010): 104–9. doi: 10.1016/j.tics.2009.12.007. Epub 2010 Feb 9. PMID: 20149715; James Dow, "Is Religion an Evolutionary Adaptation?," *Journal of Artificial Societies and Social Simulation* 11, no. 2 (2008).

3 Francis Galton, "The Part of Religion in Human Evolution," *National Review* 23 (1894): 755–63, https://galton.org/essays/1890-1899/galton-1894-religion-evolution.pdf. Accessed May 9, 2024.

4 E. O. Wilson, *Sociobiology: The Abridged Edition* (Harvard University Press, 1980), 286.

5 Paul Bloom, "Is God an Accident?," *Atlantic*, December 2005, www.theatlantic.com/magazine/archive/2005/12/is-god-an-accident/304425/. Accessed May 9, 2024.

6 "Critique Topples *Nature* Paper on Belief in Gods," *Retraction Watch*, July 7, 2021, https://retractionwatch.com/2021/07/07/critique-topples-nature-paper-on-belief-in-gods/. Accessed May 9, 2024.

7 Huda, "Creation of the Universe and Evolution in Islam," *Learn Religions*, August 27, 2020, learnreligions.com/creation-of-the-universe-2004201. Accessed May 9, 2024.

8 "The Existence of God," *Bitesize*, BBC, n.d., www.bbc.co.uk/bitesize/guides/zv2fgwx/revision/1. Accessed May 9, 2024.

9 "The Buddhist Universe," *Religions*, BBC. Last updated November 23, 2009. Accessed October 12, 2023, www.bbc.co.uk/religion/religions/buddhism/beliefs/universe_1.shtml. Accessed May 9, 2024.

10 L. A. Barnes, "The Fine-Tuning of the Universe for Intelligent Life," *Publications of the Astronomical Society of Australia* 29, no. 4 (2012): 529–64. doi:10.1071/AS12015.

11 Paul Davies, "The Anthropic Principle," *Horizon* series, BBC, Season 23, Episode 17, May 18, 1987, available at SearchBucket2, "Horizon—The Anthropic Principle—Part 2 of 4," YouTube, www.youtube.com/watch?v=r5aaBDbHl8I&t=51s. Accessed May 9, 2024.

12 Don Lincoln, "Why Does the Universe Appear Fine-Tuned for Life to Exist?," *Big Think*, October 7, 2023, https://bigthink.com/hard-science/universe-fine-tuned-life-exist/. Accessed May 9, 2024.

13 George F. R. Ellis, "The Anthropic Principle: Laws and Environments," in *The Anthropic Principle: Proceedings of the Second Venice Conference on Cosmology and Philosophy (1988)*, ed. F. Bertola and U. Curi (Cambridge University Press, 1993), 30.

Notes

14 Henry Margenau and Roy Abraham Varghese, eds., *Cosmos, Bios, Theos: Scientists Reflect on Science, God, and the Origins of the Universe, Life, and Homo Sapiens* (La Salle, IL: Open Court, 1992), chapter 16, p. 83.

15 Arno Penzias, quoted in Malcolm W. Browne, "Clues to Universe Origin Expected," *New York Times*, March 12, 1978, www.nytimes.com/1978/03/12/archives/clues-to-universe-origin-expected-the-making-of-the-universe.html. Accessed May 9, 2024.

16 Stephen Hawking and Leonard Mlodinow, *The Grand Design* (New York: Bantam Random House, 2010), 165.

17 C. S. Lewis, "Is Theology Poetry?," *They Asked for a Paper* (London: Geoffrey Bles, 1962), 211.

18 Thomas Aquinas, *God*, Book 1 of *Summa Contra Gentiles*, translated by Anton C. Pegis (1975; reprint, Notre Dame, IN: University of Notre Dame Press, 2005), 63.

19 Feser, *Aquinas* (Kindle edition, location 532).

20 Michael Egnor, "Aquinas' Fourth Way: Light in a Mirror," *Evolution News*, October 17, 2019, https://evolutionnews.org/2019/10/aquinas-fourth-way-light-in-a-mirror/. Accessed May 9, 2024.

21 M. Preiner et al., "The Future of Origin of Life Research: Bridging Decades-Old Divisions," *Life* (Basel) 10, no. 3 (February 26, 2020): 20. doi:10.3390/life10030020. PMID: 32110893; PMCID: PMC7151616.

22 D. Fox, "What Sparked the Cambrian Explosion?," *Nature* 530 (2016): 268–70. doi.org/10.1038/530268a.

23 M. P. Smith and D. A. Harper, "Causes of the Cambrian Explosion," *Science* 341, no. 6152 (September 20, 2013): 1355–56. doi:10.1126/science.1239450. PMID: 24052300.

24 Eric Hedin, "Thoughts of Evil in a Designed World," *Evolution News*, October 11, 2023, https://evolutionnews.org/2023/10/thoughts-of-evil-in-a-designed-world/. Accessed May 9, 2024.

25 Martin Schönfeld and Michael Thompson, "Kant's Philosophical Development," *Stanford Encyclopedia of Philosophy* (Winter 2019 Edition), ed. Edward N. Zalta, https://plato.stanford.edu/archives/win2019/entries/kant-development. Accessed May 9, 2024.

26 Natalie Wolchover, "Physicists Debate Hawking's Idea That the Universe Had No Beginning," *Quanta*, June 6, 2019, www.quantamagazine.org/physicists-debate-hawkings-idea-that-the-universe-had-no-beginning-20190606/. Accessed May 9, 2024.

27 David W. Deamer, "Self-Assembly Processes Were Essential for Life's Origin," *Assembling Life: How Can Life Begin on Earth and Other Habitable Planets?* (Oxford University Press, 2019; online ed., Oxford Academic, November 12, 2020). doi.org/10.1093/oso/9780190646387.003.0010. Accessed October 13, 2023.

28 Agnes Ullmann, "Louis Pasteur," *Britannica*, September 24, 2023, www.britannica.com/biography/Louis-Pasteur. Accessed May 9, 2024.

29 Ralph Lewis, "The Physical Evolution of Consciousness," *Psychology Today*, July 22, 2018, www.psychologytoday.com/ca/blog/finding-purpose/201807/the-physical-evolution-consciousness. Accessed May 9, 2024.

30 Peter Atkins, "Atheism and Science," in *The Oxford Handbook of Religion and Science*, ed. Philip Clayton (Oxford University Press, 2008; online ed., Oxford Academic, September 2, 2009). doi.org/10.1093/oxfordhb/9780199543656.003.0009. Accessed May 9, 2024.

Notes

Chapter 13: Does AI Really Change Everything? Anything?

1. Chris Vallance, "Google Engineer Says Lamda AI System May Have Its Own Feelings," BBC, June 13, 2022, www.bbc.com/news/technology-61784011. Accessed May 6, 2024.
2. Tiffany Wertheimer, "Blake Lemoine: Google Fires Engineer Who Said AI Tech Has Feelings," BBC, July 23, 2022, www.bbc.com/news/technology-62275326. Accessed May 6, 2024.
3. George Dvorsky, "Experts Sign Open Letter Slamming Europe's Proposal to Recognize Robots as Legal Persons," *Gizmodo*, April 13, 2018, https://gizmodo.com/experts-sign-open-letter-slamming-europe-s-proposal-to-1825240003. Accessed May 6, 2024.
4. David Kushner, "Discover Interview: The Man Who Builds Brains," *Discover*, February 4, 2010, www.discovermagazine.com/technology/discover-interview-the-man-who-builds-brains. Accessed July 19, 2024.
5. Antonio Regalado, "What It Will Take for Computers to Be Conscious," *Technology Review*, October 2, 2014, www.technologyreview.com/2014/10/02/171077/what-it-will-take-for-computers-to-be-conscious/. Accessed July 21, 2024
6. Closer to Truth, "Christof Koch—Is Consciousness Ultimate Reality?," YouTube, March 13, 2024, www.youtube.com/watch?v=FPMBgYzAanc. Accessed July 22, 2024.
7. Stanislas Dehaene et al., "What Is Consciousness, and Could Machines Have It?," *Science* 358 (2017): 486–92. doi:10.1126/science.aan8871. Accessed July 19, 2024.
8. Ryota Kanai, "Do You Want AI to Be Conscious?," *Nautilus*, June 9, 2021, https://nautil.us/do-you-want-ai-to-be-conscious-236578/. Accessed July 19, 2024.
9. Joshua Rothman, "Why the Godfather of A.I. Fears What He's Built," *New Yorker*, November 13, 2023, www.newyorker.com/magazine/2023/11/20/geoffrey-hinton-profile-ai. Accessed July 19, 2024.
10. Steven Levy. "If Ray Kurzweil Is Right (Again), You'll Meet His Immortal Soul in the Cloud," *Wired*, June 13, 2024, www.wired.com/story/big-interview-ray-kurzweil/. Accessed July 19, 2024.
11. Simonne Shah, "Sam Altman on OpenAI, Future Risks and Rewards, and Artificial General Intelligence," *Time*, December 12, 2023, https://time.com/6344160/a-year-in-time-ceo-interview-sam-altman/. Accessed July 19, 2024.
12. Creative Destruction Lab, "Geoff Hinton: On Radiologists," YouTube (at 0:32), www.youtube.com/watch?v=2HMPRXstSvQ. Accessed July 18, 2024.
13. James M. Milburn, "How Will We Solve Our Radiology Workforce Shortage?," *Bulletin*, March 1, 2024, www.acr.org/Practice-Management-Quality-Informatics/ACR-Bulletin/Articles/March-2024/How-Will-We-Solve-Our-Radiology-Workforce-Shortage. Accessed July 18, 2024.
14. Jason Arunn Murugesu, "How Newly Discovered Brain Cells Have Made Us Rethink the Human Mind," *New Scientist*, February 19, 2024, www.newscientist.com/article/mg26134791-500-how-newly-discovered-brain-cells-have-made-us-rethink-the-human-mind/. Accessed July 21, 2024.
15. Quoted in Gareth Cook, "Does Consciousness Pervade the Universe?," *Scientific American*, January 14, 2020, www.scientificamerican.com/article/does-consciousness-pervade-the-universe/. Accessed July 16, 2024.
16. Philip Goff, "Consciousness: Why a Leading Theory Has Been Branded

Notes

'Pseudoscience,'" *The Conversation*, September 29, 2023, https://theconversation.com/consciousness-why-a-leading-theory-has-been-branded-pseudoscience-214214. Accessed July 16, 2024.

17. Dan Falk, "Is Your Brain a Computer?," *Technology Review*, August 25, 2021, www.technologyreview.com/2021/08/25/1030861/is-human-brain-computer/. Accessed May 6, 2024.

18. Eric Holloway, "Do Chatbots Really Understand Things," *Mind Matters News*, July 20, 2024, https://mindmatters.ai/2024/07/tech-hype-watch-do-chatbots-really-understand-things/. Accessed July 20, 2024.

19. Robert J. Marks II, "Things Exist That Are Unknowable," *Mind Matters News*, March 19, 2019, https://mindmatters.ai/2019/03/things-exist-that-are-unknowable/. Accessed July 24, 2024. See also Manon Bischoff, "These Are the Most Bizarre Numbers in the Universe," *Scientific American*, May 23, 2023, www.scientificamerican.com/article/these-are-the-most-bizarre-numbers-in-the-universe/. Accessed July 24, 2024.

20. "Why Human Creativity Is Not Computable," *Mind Matters News*, March 30, 2021, https://mindmatters.ai/2021/03/why-human-creativity-is-not-computable/. Accessed July 24, 2024.

21. "The Paradox of the Smallest Uninteresting Number," *Mind Matters News*, March 29, 2021, https://mindmatters.ai/2021/03/why-all-positive-integers-are-interesting-proof-by-contradiction/. Accessed July 24, 2024.

22. Robert J. Marks, "The Software of the Gaps: An Excerpt from Non-Computable You," *Mind Matters News*, June 24, 2022, https://mindmatters.ai/2022/06/the-software-of-the-gaps-an-excerpt-from-non-computable-you/. Accessed June 24, 2024.

23. Marks, "The Software of the Gaps."

24. Jeffrey Lee Funk and Gary N. Smith, "Delivery Drones, Robotaxis, Even Insurance—Wildly Hyped Dreams for AI Startups Are Giving Tech Investors Nightmares," *MarketWatch*, June 13, 2022, www.marketwatch.com/story/delivery-drones-robotaxis-even-insurance-wildly-hyped-dreams-for-ai-startups-are-giving-tech-investors-nightmares-11655111226. Accessed July 24, 2024.

25. Robert J. Marks, "AI Is No Match for Ambiguity," *Mind Matters News*, June 17, 2019, https://mindmatters.ai/2019/06/ai-is-no-match-for-ambiguity/. Accessed May 6, 2024.

26. Paul C. Allen, "The Allen Institute for Artificial Intelligence to Pursue Common Sense for AI," Cision PR Newswire, February 28, 2018, www.prnewswire.com/news-releases/the-allen-institute-for-artificial-intelligence-to-pursue-common-sense-for-ai-300605609.html. Accessed May 6, 2024.

27. Robert J. Marks, "New AI Learns to Simulate Common Sense," *Mind Matters News*, December 30, 2021, https://mindmatters.ai/2021/12/new-ai-learns-to-simulate-common-sense/. Accessed May 6, 2024.

28. Gary Smith, "The Great American Novel Will Not Be Written by a Computer," *Mind Matters News*, June 30, 2021, https://mindmatters.ai/2021/06/the-great-american-novel-will-not-be-written-by-a-computer/. Accessed May 6, 2024.

29. Noor al-Sabai, "ChatGPT's Dirty Secret: It's Powered by 'Grunts' Making $15 Per Hour," *Futurism*, May 10, 2023, https://futurism.com/the-byte/chatgpt-15-hour-workers. Accessed May 6, 2024.

30. Sara Goudarzi, "Popping the Chatbot Hype Balloon," *Bulletin of the Atomic*

Notes

Scientists, September 11, 2023, https://thebulletin.org/premium/2023-09/popping-the-chatbot-hype-balloon/#post-heading. Accessed July 18, 2024.

31 Robert J. Marks, "Did the GPT3 Chatbot Pass the Lovelace Creativity Test?," *Mind Matters News*, December 14, 2022, https://mindmatters.ai/2022/12/did-the-gpt3-chatbot-pass-the-lovelace-test/. Accessed May 6, 2024.

32 Gary Smith, "LLMs Can't Be Trusted for Financial Advice," *Journal of Financial Planning*, May 2024, www.financialplanningassociation.org/learning/publications/journal/MAY24-llms-cant-be-trusted-financial-advice-OPEN. Accessed July 22, 2024.

33 Gary Smith, "Internet Pollution: If You Tell a Lie Long Enough...," *Mind Matters News*, https://mindmatters.ai/2024/01/internet-pollution-if-you-tell-a-lie-long-enough/. Accessed July 22, 2024.

34 Sara Goudarzi, "Popping the Chatbot Hype Balloon," *Bulletin of the Atomic Scientists*, September 11, 2023, https://thebulletin.org/premium/2023-09/popping-the-chatbot-hype-balloon/#post-heading. Accessed July 18, 2024.

35 Smith, "Internet Pollution."

36 Ben Lutkevitch, "Model Collapse Explained: How Synthetic Training Data Breaks AI," TechTarget, July 7, 2023, www.techtarget.com/whatis/feature/Model-collapse-explained-How-synthetic-training-data-breaks-AI. Accessed July 23, 2024.

37 Darren Orf, "A New Study Says AI Is Eating Its Own Tail," *Popular Mechanics*, October 20, 2023, www.popularmechanics.com/technology/a44675279/ai-content-model-collapse/. Accessed July 23, 2024; I. Shumailov et al., "The Curse of Recursion: Training on Generated Data Makes Models Forget," 2023, ArXiv, abs/2305.17493.

38 John Horgan, "Will Artificial Intelligence Ever Live Up to Its Hype?," *Scientific American*, December 4, 2020, www.scientificamerican.com/article/will-artificial-intelligence-ever-live-up-to-its-hype/. Accessed May 6, 2024.

Conclusion: The Truths That Matter Most

1 A. J. Bergesen, "Political Witch Hunts: The Sacred and the Subversive in Cross-National Perspective," *American Sociological Review* 42, no. 2 (1977): 220–33. doi.org/10.2307/2094602.

2 "Frederick Engels' Speech at the Grave of Karl Marx, Highgate Cemetery, London, March 17, 1883," transcribed by Mike Lepore, 1993, www.marxists.org/archive/marx/works/1883/death/burial.htm. Accessed April 4, 2024.

3 William Provine, "Darwinism Science or Naturalistic Philosophy?," *Origins Research* 16, no. 2 (1994): 9, www.arn.org/orpages/or161.htm. Accessed May 9, 2024.

About the Authors

Dr. Michael Egnor is Professor of Neurosurgery and Pediatrics and Neurosurgery Residency Director at the Renaissance School of Medicine in Stony Brook, New York. He is also a Senior Fellow at the Discovery Institute's Center for Natural and Artificial Intelligence.

Denyse O'Leary is the author or coauthor of several books on the topics of science, creation, design, and spirituality, including *The Spiritual Brain: A Neuroscientist's Case for the Existence of the Soul* (coauthored by neuroscientist Mario Beauregard).